The C

The Octavo
(Roundworld Edition)
A Sorcerer-Scientist's Grimoire.-

Published by
Mandrake of Oxford
PO Box 250
OXFORD
OX1 1AP (UK)

A CIP catalogue record for this book is available from the British
Library and the US Library of Congress.

Contents

Chapter 0
Introduction

Every generation rewrites its theory and practise of Magic. Chaos Magic updated the ancient Occult Arte during the last 2 decades of the 20th century. It liberated magic from its dependency on religious symbolism and theological theories about deities and spirits by demonstrating that imaginary gods and spirits have exactly the same effects as the supposedly 'real' ones. Thus it initiated the reformulation of magical theory and practise in terms of some of the symbolism of science, psychology, and parapsychology. Plus its own idiosyncratic symbolism went down rather well with the late crop of hippies and then the punks. The eight-rayed Chaostar looks particularly catching in chunky silver jewellery or in red paint on black leather. Today a lot of them wear a discreet Octaris ring with a business suit.

However, during periods of improved sanitation and reduced warfare, some generations even live long enough to try and rewrite magic one and-a-half times.
Thus the old magus Stokastikos presents yet another thesis, this Octavo, perhaps his last after nearly 40 years spent slaving over a hot pentacle.

Now reality imitates art, because art explores the imagination, and we create reality mainly through imagination, (or lack of it).

Paracelcus, Eliphas Levi, MacGregor Mathers, Aleister Crowley, Austin Spare, and Michael Moorcock all fed ideas into Chaos Magic. Plus it made some acknowledgement to the ideas of Quantum Physics and other bits of strange science.

As the new tradition has matured it has absorbed other elements as well, and the time now seems ripe to add a further portion of Physics, (because nobody with half a brain in the developed world really believes in religion anymore), plus some more Pratchett.

Chaoists have already appropriated Sir Terry Pratchett's concept of Octarine and they have found it to have such exceptional magical efficacy that they now scour his Discworld novels for further joys and metaphysical revelations.

It will perhaps turn out that Pratchett's Discworld can tell us as much about our world's magic as Edwin Abbot's Flatland told us about our higher dimensional geometry and sociology.

With the mathematicians Ian Stewart and Jack Cohen, Terry Pratchett has also authored three Science of Discworld books that not only further confirm his genius and erudition but also portray physics and science as enlightening, funny, and not as horribly difficult as it looks.

The Science of Discworld series uses the idea that only magic works on Discworld, and that only science works on Roundworld (that means here). This didactic device works well enough to explain lots of things, but it looks suspiciously like a bit of a rough approximation. In practise some science does work on Discworld, it doesn't all function on the basis of randomness, whim and instantly materialised wishes. On the other hand this world doesn't run entirely on science either. It contains an element of randomness, whim, and instantly materialised wishes - but a rather smaller one.

We should perhaps have expected something like this because Discworld has slightly more than two dimensions, it has some thickness as well, and our universe seems to have slightly more than three dimensions, making it hyperspherical.

Plus this world has a much higher value of lightspeed and a much lower value of Planck's constant than Discworld, so it takes considerably more effort of sorcery here to overcome the stricter causality.

Discworld and Roundworld both contain legions of imaginary gods that have real effects. This should not surprise us, both contain people as well and they provide excellent hosts for such deities.

The wizards of Discworld refer to the ill-understood science of their realm as "that err, quantum stuff". In this Roundworld it seems that similar "err, quantum stuff" probably accounts for much of its ill-understood magic.

Discworld contains a Mighty Grimoire called The Octavo, the creation manual for the Discworld itself, lying hidden in the vaults of its Unseen University.

When asked during an invocation if this world conceals its own version, the goddess Apophenia, (patroness and muse of unexpected ideas) cackled and replied: -

'Well of course it will, but don't confuse it with that Necronomicon which fell in from elsewhere and partially disintegrated on impact. A much weirder and far more frightening edition applies to this world.'

This reconstructed Roundworld Edition of the Octavo comes from fragments discovered in various Chaoist classics and in the vaults and minds of Arcanorium College, together with sly hints from Apophenia herself.

It of course consists of Eight chapters, each prefaced with one of the great spells that appear to structure this reality. To dissuade casual enquiry, and for Health and Safety reasons, we have expressed these spells tersely in rigorous mathematical form, but then we have relented somewhat and provided a written exposition as well, together with example practical applications for using each of them.

The Magical Ritual develops as we go along from centring, to encircling, to adding deities, and then via chaometry and probability distortion considerations and the magical link, to parapsychological action. Finally we give some consideration to the consequences of magical action, which err, perhaps we should have thought about first. So read the whole grimoire before trying anything.............

Please avoid collateral damage. We inhabit a rather small planet.

Stokastikos, Past Grandmaster of the Magical Pact of the Illuminates of Thanateros, Chancellor of Arcanorium College.

Peter J Carroll. Albion Southwest. 2010

Dedications.

Sir Terence David John Pratchett, (1948 -) Novelist, and to some, a philosopher of considerable genius.

'The truth may be out there, but the lies are inside your head.'
Terry Pratchett, *"Hogfather", footnote.*

'Seeing, contrary to popular wisdom, isn't believing. It's where belief stops because it isn't needed anymore.' Terry Pratchett, *Pyramids.*

(Sir Terry's brilliant Discworld novels treat the subject of magic with a humorous, scholarly, humanistic, eclectic irreverence that provides a splendidly entertaining inspiration to all aspiring magicians; so read the whole lot and learn.)

Hypatia of Alexandria. Mid 4th Century – 415AD. Neo-platonist pagan philosopher, mathematician and astronomer. Her murder marks the beginning of the dark ages.

Emmy Noether. 1882 – 1935. Abstract algebraist and theoretical physicist. Her theorem explains the fundamental connection between symmetry and conservation laws.

Gottfried Wilhelm von Leibniz, (1646 - 1716) Polymath, mathematician and philosopher.

``Sans les mathématiques on ne pénétre point au fond de la philosophie.
Sans la philosophie on ne pénétre point au fond des mathématiques.
Sans les deux on ne pénétre au fond de rien."

Or, liberally translated;

'Without mathematics (some theory of 'how much') you cannot understand philosophy.
Without philosophy (some theory of 'how') you cannot understand anything'.

(Quantitative analysis thus elucidates Qualitative analysis, and we apologise in advance for any algebraic difficulties that you may encounter, but at root mathematics merely involves a consideration of 'how much', a concept which even Off-Piste Physicists cannot ignore).

The Staff and Alumni of Arcanorium College.

In the best traditions of Invisible Colleges and Unseen Universities, the conspirators of Arcanorium meet in two bomb proof vaults, one beneath Southern England, the other somewhere beneath Southern California. www.arcanoriumcollege.com. My thanks for their insights, inputs, and illustrations. Particular thanks go to Chaosage, RinceWIT, Paul, and Milodonk for editorial suggestions, and exceptional thanks to Jo and Matt Kaybryn for both editorial suggestions and sumptuous illustration and cover design.

The Publisher. Hurrah for Mogg Morgan of Mandrake, for his superhuman dedication and patience and attention to detail in maintaining the finest traditions of esoteric publication.

The Octavo – The Grimoire

The Spells of the Binding.

$$\frac{M}{L} = \frac{c^2}{G} \qquad\qquad \frac{GM}{L^2} = A$$

2) The Spells of the Spinning.

$$W^2 = \frac{2\pi GM}{V_H} \quad \text{(Vorticitation)}$$

$$f = \frac{c}{2L} \quad \text{(Frequency)}$$

3) The Spells of Illusion.

$$z = \frac{\lambda_o}{\lambda_e} - 1 = \frac{GM}{c^2} \frac{1}{L-a} - 1 \quad \text{(Redshift.)}$$

$$L = 1 + \sqrt{d - d^2} - d \quad \text{(Hyperspherical lensing)}$$

4) The Spells of Subtle Magic.

$$\Delta E \Delta t \sim \hbar \sqrt[3]{U} \qquad \Delta S K^0 \, \Delta t \sim \hbar \sqrt[3]{U} \qquad \Delta H k K^0 \, \Delta t \sim \hbar \sqrt[3]{U}$$
(Energy)   (Entropy)   (Information)

5) The Spells of the Linking.

$$d = \sqrt{s^2 - (ct)^2 + (ct_i)^2}$$
(3 dimensional time, compact form)

$$0 = \sqrt{s^2 - (ct)^2} \quad \text{(Entanglement)}$$

$$0 = \sqrt{(ct_i)^2 - (ct)^2} \quad \text{(Superposition)}$$

6) The Spells of Impractical Magic.

$$P_\Psi = P + (1\text{-}P)\,\Psi^{1/P} \qquad\qquad P_\Psi = P\text{-}P\,\Psi^{1/(1-P)}$$

(Spell)   (Antispell)

7) The Spell of Practical Magic.

$$\Psi = GLSB$$

8) The Spell of the Narration.

$$U^2 = H = \Phi + S/k$$

1. The Spells of the Binding

$$\frac{M}{L} = \frac{c^2}{G} \qquad\qquad \frac{GM}{L^2} = A$$

For millennia magicians, philosophers and scientists and various other explorers have sought The Map of reality. This map has grown exponentially larger with the passage of time.

In the first aeon[1] the ancient Shamans had maps that extended no further than they could walk, no further up than what they could see in the sky at night, and no further down than their deepest caves. It extended no further metaphysically than what they could observe and intuit from various natural phenomena including their semi-articulated experience of themselves. Nevertheless they did manage to forge workable systems of science (simple technology), religion, art, and magic using these maps. Shamanic magic still works reasonably well, especially for line of sight work, magic within the village or hunting grounds, and occasionally for magic just over the perceptual horizon and into the imagination, like tomorrows weather or battle.

In the second aeon our pagan ancestors built slightly better maps of reality. The maths and the astronomy helped a bit, but much of the growth of The Map came in the dimension of what we now call psychology. Arguably we have made very little progress at all in our psychological understanding of ourselves since the classical Greeks put down what they knew in terms of mythological stories and in their philosophies. An education in the classics still illuminates as many of the persistent human verities as does any contemporary psychology course. Organised paganism developed along with agriculture and the founding of city-states, and in pagan magic we begin to see a shift of emphasis from direct magical interaction with the immediate environment towards wider goals. Such wider goals included the fates of the city-states themselves and the psychological expansion of individuals via the activities of numerous mystery schools.

The collapse of classical paganism owed much to the crisis of classical civilisation itself. As empires replaced the city-state, the new empires could scarcely manage to accommodate all the conflicting pagan philosophies, customs, and moralities within them, and thus they mostly adopted a simple and bloody-minded solution, monotheism. One leader, one priesthood, one set of rules. Obey or die.

The third aeon of monotheism brought an expansion of the geographical horizons of the map of reality but a contraction in its psychological dimensions. Whilst exploration and conquest opened up new continents, self-knowledge actually contracted and inner horizons shrank under totalitarian religious structures. Such structures also inhibited the cosmological expansion of the map, and the universe remained modelled at about the same size for several thousand years. Magic made little progress during the monotheist aeon. The monotheist religions appropriated certain magical ideas into their creeds and rituals and forbade all other magic on pain of death. The Christian ideas of icons, relics and transubstantiated hosts all devolve from neo-platonic style magical thought.

During these dark ages The Map turned monochrome. Everything devolved into a simple battle between good and evil. The grimoires that survive from the monotheist period mostly seem rather dull affairs. The wizard conjures various spirits either in the name of deity or in the name of the anti-deity or devil, and mostly for fairly mundane results magic effects such as seduction, wealth, divination and the destruction of rivals. Ideas about secret wisdom, psychological development and mystical inspiration seem to have got lost.

In the dark ages of the third aeon the monotheists merely added Hell to complement Heaven and Earth.

Towards the Renaissance however we do observe some minor improvements with the publication of Abramelin [2] and the flowering of Kaballah, yet how primitive the Kabbalistic Tree of Life looks now, it encompasses little more than the planets known to the ancients and a bit of pagan psychological understanding overlaid with a monotheist gloss.

The fourth aeon of atheism began with a restart of the rational enquiry into the natural world that the pagans had originally initiated, and a rational enquiry into the nature of religion initiated by more Protestant minded monotheists. Within a few centuries it had added huge numbers of orders of magnitude to the physical macrocosm and microcosm of The Map. Quite suddenly humanity discovered an awesomely vast universe beyond the solar system and an incredibly small and intricate structure in all its parts. Rather quickly it became apparent that the monotheist god as described in various sacred texts seemed a hopelessly small and silly idea with which to explain the vast new map. Some like Liebniz, Spinoza and Einstein began to regard the monotheist god as little more than a primitive abstract metaphor for the universe itself. Those who wished to continue believing in this deity as a personal one with a human personality invested much effort in denying or selectively ignoring the size of the new Map.

Since perhaps late classical antiquity The Map had two principle domains, the Macrocosm and the Microcosm, the heavens and the world of mankind. The legendary Emerald

Tablet of Hermes Trismegistus famously proclaimed 'as above, so below', asserting some form of symmetry between the heavens and the world of humanity. Hermeticism actually now appears to have consisted of an attempt to translate Egyptian magical texts into Greek, replete with inevitable translation errors and a Hermes Thrice-Magus who probably consisted of a whole school of first and second century esotericists.

At this point in the exegesis of The Map we note that the aeons denote the progression of ideas and cultural paradigms rather than precise historical periods. At present we still have widespread remnants of third aeon activity despite that fourth aeon ideas dominate in leading cultures, and fifth aeon ideas begin to manifest.

Unsurprisingly the atheistic aeon brought a fresh flowering of magical thought. Religion lost much of its power to persecute magicians and science contented itself with merely disparaging magic whilst occasionally dabbling in a bit of parapsychological research and some rudimentary investigations into the power of belief and suggestion. Meanwhile the decline of conventional authoritarian religion brought about a renaissance of popular interest in esoteric subjects.

Towards the end of the third aeon and throughout the fourth we discovered that the macrocosm has an almost unimaginably vast size and that the microcosm extended a very long way below what the naked eye could perceive. Our sun stands revealed as merely one of billions of stars in the home galaxy and galaxies in their billions populate the universe. Various types of microscope revealed first the bacteria and biological cells and then atoms and then finally fundamental particles.

18

Thus we now need to consider a map with three main size perspectives:

The Macrocosm: 10^7 to 10^{26} metres, from planet size to the limits of the universe.

The Midicosm: 10^{-5} to 10^7 metres, the smallest stuff visible to the naked eye up to planet size.

The Microcosm: 10^{-5} to 10^{-35} metres, from bacteria down to the quantum scale.

(A brief note now follows on index notation; this saves a lot of mental pain and paper.

$4 \times 10^{-6} = 0.000,004$, four millionths.

$3.5 \times 10^9 = 3,500,000,000$, three and a half billion.

3.5×10^9 times $2 \times 10^{20} = 7 \times 10^{29}$, a number too big to have a simple name.

With indices you just multiply the numerical part and add the indices to multiply, or do the reverse of this to divide.)

The fifth pandaemon-aeon begins where the atheistic aeon finishes in its late nihilistic phase, with The Death of Certainty. Cultural relativism undermined the claims of absolute value systems, making all values situational. Neurophysiology negated the post-monotheist misconception of a unitary self, replacing it with a neo-pagan multimind model. Quantum physics negated causality, replacing it with a microcosm that runs on chaos and chance. On the macroscopic scale we seem to have a causality problem concerning the very existence of the cosmos. The fourth aeon Big Bang hypothesis cannot explain what causes the universe to exist unless it resorts to some sort of creation ex-nihilo idea; it merely hypothesises a series of changes from a previous state, and a fantastically improbable prior one at that.

As some parts of humanity enter the fifth aeon, The Map has expanded enormously but much of the cartography of the previous aeons now appears rather sloppy. In particular the third aeon God hypothesis and the fourth aeon Causality hypothesis now look rather like wishful thinking.

Whilst the second aeon pagans probably managed to explore near to the limits of human psychology, the fourth aeon scientists have probably got fairly close to the physical limits of the Map. We probably now know the exact size of the universe and the size of its smallest possible components. However, much remains to explore, unimaginable technologies with which we can modify our environment and ourselves await discovery no doubt, and we have barely started on a systematic exploration of magic.

Knowing the physical size and shape of the territory makes for a much better map. We abandoned flat earth theory some time ago, for a second time. The ancient Greeks had abandoned it previously, but then we had it rekindled in the dark monotheist aeon. A poor map fails to show the full relationships between the elements on it, like the existence of America or how to get from there straight to East Asia. Now we seem poised to abandon flat universe theory in favour of something more useful which also tells us something more about the relationships of its parts.

Flat universe theory states that although gravity curves spacetime, the universe has no overall curvature because the gravity associated with all the mass in it exactly cancels out against the apparent expansion of the universe which began with a big bang. Cosmologists now struggle to incorporate an ever increasing number of apparently contradictory observations into this creaking hypothesis. Flat universe theory contains multiple inelegancies. It now asserts that absolutely everything must have erupted from a singularity of zero size and infinite density and then gone on to expand at a first decreasing and then an accelerating rate, assisted by various forms of currently imperceptible dark matter and dark energy which somehow account for the inability of the matter that we can observe to yield the universe that we do observe.

In short it all resembles an increasingly grotesque mess, like a lot of those medieval maps with phlogiston[3], improbable sea monsters, speculative continents, and 'edge of the world' marked on them.

The replacement of flat universe theory with hypersphere theory seems immanent if not overdue, and the master equation for hypersphere theory looks like this:

The Spell of the Binding: $$\frac{M}{L} = \frac{c^2}{G}$$

It also has a Corollary: $$\frac{GM}{L^2} = A$$

This simple equation states that the mass M, divided by the length L, (to the antipode, the furthest distance in the universe) equals lightspeed c, squared, divided by the gravitational constant G.

Miraculously this equation also applies to every single fundamental matter particle within the universe as well, (where L then shows the wavelength). Thus it unites the microcosm and the macrocosm.

We call this The Spell of The Binding because it holds everything together, quite literally. In chapter 2 we shall examine the complementary Spell of the Spinning which stops everything from collapsing into itself. (The corollary allows us to measure the exact size of the universe). It does seem that the four elements of this spell do not merely have a coincidental relationship but rather that any three of them together defines the value of the fourth, if any of them changed it would presumably result in the change in the value of at least one of the others.

The relevance of this to magical theory and practise will hopefully appear clearly over the next several chapters which make the case for a three dimensional time which supplies the medium for magic and which older Grimoires refer to rather imprecisely as 'higher dimensions' or 'the astral planes'.

This equation may not look much like a map, but it does indicate the size and shape of the universe, and the shape of all the individual matter particles within it. This shape comes out as a simple hypersphere, or what mathematicians call a 3-Sphere or a 4-Ball, and we may have to use some tricks to render it possible to visualise.

Well you didn't expect the universe to resemble a simple ball or a shapeless infinity did you? Neither of those makes any sense. Any simple ball would have an outside and thus it could not constitute the entire universe itself. Infinity makes no sense either, the concept only arises when we make the mistake of dividing by zero to get a silly answer. The hyperspherical universe actually has finite and unbounded extent in both space and time. This means that a strictly limited (although vast) amount of each exists and you can in principle travel as far round it in space and time as you like without encountering any boundaries but you will eventually get back to either 'where' OR 'when' you started from. However you cannot get back to both 'where' AND 'when' you started from, thus you cannot create a paradox by attempting to assassinate your own grandfather.

Down at the microcosmic end of reality, on the quantum scale, we encounter similar difficulties in visualising the structure. Fundamental particles cannot consist of zero size points in space and time or else all sorts of ridiculous infinities would occur which we do not actually observe. Instead a certain minimum size related to Planck's constant applies, and nothing gets any smaller than this. At the Planck scale fundamental particles seem to have the same sort of geometry and topology as the universe itself. They consist of hyperspheres as well, objects which curve back in on themselves so that nothing inside can get out without falling back in, due to their intense spacetime curvature. However unlike the universe itself the quantum hyperspheres do not have enough room inside to permit anything from the outside to fall in. The universe never faces the possibility of stuff falling into it, so far as we know, because no space or time exists outside of it to contain anything else. The mass and energy within it subtends all the space and time which exists.

Thus Hermes Trismegistus et al, seem to have intuited the situation correctly, a deep symmetry does exist between the macrocosm and the microcosm. However we need to look to the far reaches of both to see it, and when we do we cannot 'see' it in any optical sense, although the mathematical symmetries reveal themselves. Meanwhile back in the midicosm of human sized phenomena we remain subject to the strange and subtle if not downright occult effects of the structure of the macrocosm above and the microcosm below. In particular these give rise to the non-local and the random aspects of reality respectively. These non-local and random aspects of reality create some peculiar and newly recognised scientific effects; and they also lie at the root of what we have for millennia called magic.

Before considering these effects we should perhaps pause and consider the geometry and topology of the hyperspheres that seem to lie at the macrocosmic and microcosmic extremities of The Map.

An ordinary sphere has a finite but unbounded two dimensional surface. This surface has no edges and you can keep going round it for any distance.

The basic hypersphere (or 3-sphere or 4-ball as mathematicians call it), differs from an ordinary sphere (or 2-sphere or 3-ball) in that it closes itself in three dimensional space by having an escape velocity greater than the maximum possible velocity of lightspeed. This effectively closes space around it so that nothing can get out. Gravity bends space in all three of the ordinary directions at once, gravity acts as the fourth dimension about which the other three bend. In some ways this resembles what we call a black hole but worse, hyperspheres have escape velocities of root 2 lightspeed, about 1.414c.

Thus nothing inside can get out, and if anything tries it simply follows the spacetime curvature and finds itself returning to where it started from. Imagine in the case of the universe itself, a giant fishbowl with walls made of gravity. Imagine in the case of a fundamental particle something similar but containing only enough room for a single tightly enclosed particle and that every particle has one of these tight fitting fishbowl suits, in fact it consists of it.

Now we cannot easily visualise a hypersphere, partly because it has no edges to define it in our imaginations. Yet we can fairly easily visualise a hypercube that represents a cube closed by its spacetime curvature. We can make a two dimensional drawing of a cube by drawing two squares and joining them at their vertices (corners) to make a perspective representation of a three dimensional cube. See figure 1. Now using the same scheme we can make a three dimensional perspective model of a hypercube by simply using two cubes joined at their vertices. In lieu of actually supplying the self-assembly kit for this three dimensional perspective model we simply provide a perspective diagram of the perspective model, (get your own cocktail sticks and glue).

Fig 1. Hypercube

Consider the hypercube representation, and remember that it shows a hypercube in perspective with necessarily distorted angles and lengths. The six truncated pyramidal shapes on the faces of the central cube actually represent cubes themselves. So a 'straight' trajectory from one of these pyramids into another takes you on a circular journey through four of them back to where you started. If you attempt to leave the structure entirely you merely re-enter it on the face opposite to the one you tried to leave from, as opposite faces of the entire structure lie in contact with each other. The cube style model does not show that, and for this we need to resort to the Torus.

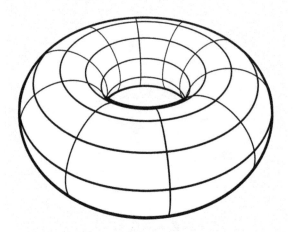

Fig 2. Torus

You can get rid of two edges of a square of paper by rolling it into a tube and joining two opposite edges. Then you can get rid of the other two edges by bending the resulting tube into a doughnut and sticking the two ends together. This constitutes a torus. Such a toroidal structure has no edges but its three dimensional equivalent does not have quite the same simple topology as a hypersphere because of the hole in the middle of the doughnut*.

Now imagine making a torus with a square cross section by folding the paper into a box section instead of rolling it into a tube. Then bend it round so that the ends of the thing approach. Then imagine using three of these square section toruses to connect opposite outer faces of the hypercube. This gives an image of the actual topology of a hypercube by showing its self-connectedness.

*Poincare[4] conjectured in 1904 the fairly obvious intuition that the hypsersphere constituted the simplest finite and unbounded closed shape, simpler for example than the torus. Grigori Perelman[5] proved this conjecture only in 2003, and the proof runs to hundreds of pages of challenging mathematics. It seems reasonable to conjecture further that nature will have used the simplest possible method of creating a finite and unbounded space. Nevertheless the hypercube version of it will prove useful when we come to visualise its rotation in chapter 2.

Figure 3. The Triple Torus

The Spell of the Binding has a corollary:

$$\frac{GM}{L^2} = A$$

This states that the acceleration A, arises naturally within the universal hypersphere because of its gravity/spacetime curvature. This acceleration has a negative value, it slows things down. It appears as the Anderson[6] deceleration which we have observed slowing down our deep-space probes by a tiny but measurable amount.

The acceleration that the binding spell removes form the universe does not just disappear. It reappears as orbital velocity where it noticeably affects the rotations of very large objects like galaxies. Galaxies rotate rather faster than Newtonian theories say they should and this has led to the hypothesis of a phlogiston like substance called 'dark matter' to balance the equations. However dark matter becomes quite unnecessary if we simply add the acceleration back into the orbital velocity formula: -

$$V_0 = \sqrt{Gm/r} \quad \text{thus changes to} \quad V_0 = \sqrt{Gm/r + rA}$$

This makes very little difference at planetary distances although minor anomalies in satellite slingshot manoeuvres have started to show up, but it does solve the major anomalies of galactic rotation without dark matter. Dark matter does not seem half as interesting as it sounds anyway. It would have to do absolutely nothing except make a bit of extra gravity.

Using the value of this acceleration A, we can easily calculate the mass, length, and time horizon of the universal hypersphere. The values for these quantities **all** come out as **the same** vast multiple of the Planck mass, length, and time!

$$M/m_p = L/l_p = T/t_p = U = \sim 6 \times 10^{60}$$

If $M/L = c^2/G$, then the ratios of all the Planck quantities to the Cosmic quantities, will have the same ratio, whatever the measured value of A, the Anderson acceleration. The current best estimate of A gives a ratio U, the ubiquity constant, of about 10^{60}. We have no idea of how we come to inhabit a universe of this apparent size. Nothing in our research so far shows anything that acts to phenomenise it at this size, as you will see in a later chapter we've used the same equations to calculate the hypothetical ubiquity constant of Discworld for a bit of light relief.

We cannot prove that $M/L = c^2/G$, because vast amounts of uncertainty and argument rage all over the cosmological measurement data and its interpretation.

However we intuit that it does apply exactly, largely because it gives reasonable values for M and L from the measured value of A, and it would also produce a cosmic hypersphere and quantum hyperspheres working on the same principles, and also because it has an elegant simplicity and beauty, and not least because it yields eventually a realistic model for magic as well.

It may seem a fantastically improbable coincidence that intelligent human observers just happened to appear when these ratios all appear to have roughly the same value. It seems much more likely that they have exactly the same value that the universe always has this size and that it has never expanded from some sort of singularity and big-bang. If a universe of constant mass had expanded the value of M/L would not remain constant.

The vast dimensionless number U, which equals about 6 x 10^{60}, or 6 followed by sixty noughts, shows the mass, length, and time dimensions of the universal hypersphere as sixty orders of magnitude greater than that of the microcosmic hyperspheres of its constituent fundamental particles.

Amusingly the letter U stands directly opposite h in the English alphabet arranged in a circle. We use h to represent Planck's constant for the smallest phenomena, whilst U represents the largest. The art of magic lies in contriving coincidences.

We can call this number U, 'Ubiquity', or The Ubiquity Constant, for it applies everywhere and every-when, and as later chapters of this Grimoire will show, it also intriguingly defines the amount of Information and the amount of Chaos in the universe, and it also defines the basic units of magic, the Thaum and the Prime.

So, in summary of The Map so far:

We appear to inhabit a vast but finite and unbounded space that has a slightly more complex geometry than the fairly flat Euclidian (3D) space to which we have become accustomed in the midicosm, and this makes it a little difficult to visualise. The particles which exist within this universe and which make up all the structures in it, including us, also seem to share this odd geometry. Later chapters will seek to show that the structure and dynamic behaviour of the hyperspheres at both the macroscopic and the microscopic ends of the scale have a direct effect on what can happen in the midicosmic realm.

It may not at first glance look much like a map. However you cannot really even make an ordinary map of the solar system because the bits keep moving around all the time. The only useful map of the solar system consists of the equations of its planetary motions. Any advanced map has to show the dynamic components of a system. An advanced map of the earth will show the tectonic drift of the continents. A really advanced map of the human mind would also show its evolution and development and where it could go next.

As Above, So Below, and, Also In-Between!

(The Hermeticists didn't really know about the existence of the hidden microcosm below, but now we can make inroads into the Tris-Magistery that they sought.)

Magical Ritual 1.

This work of practical magic develops along with the exegesis of the eight spells. It resumes all the techniques on which the aspiring magician can build a structure which will suffice for daily practise or for more elaborate works of magic of any kind. Where necessary the magician can shorten it for rapid magical action. No part of it remains indispensable.

It does not depend on any ancient or supposedly sacred symbolism; the rubric shows how chaos magicians develop such things to suit themselves. The symbolic systems that magicians use function rather like the symbols used in algebra. It doesn't really matter which letters of the alphabet or even which alphabets you use to represent concepts, only the meaning of the concepts, the relationships between the concepts actually matter.

Below we present the central concepts along with some symbolic representations drawn for the chaoist corpus, but basically the magician can adapt symbolism to taste and personal meaning.

Magical Ritual consists of 5 main conceptual parts:

Centring. The magician powers up for magical action.

Encircling. Connecting with the macrocosm.

Invocation. Summoning useful sources of power and knowledge.

Conjuration. Doing the magic, enchantment, divination, evocation, illumination, etc. This part can get a bit lengthy, depending on the desired objective.

Banishing. This part does not actually get rid of anything; it just sets the magic free to do as the magician has specified.

Decades of painstaking research have shown that the most effective practical magic involves all five of these procedures, despite that the literature and anecdotal history of the subject lies littered with occasional impressive or lucky results arising where shortcuts have occurred due to laziness, ignorance, or emergency circumstances. However for consistently useful results, aspiring magicians should learn the full technique.

Chapters 1, 2, and 3 relate to the first 3 parts of the Magical Ritual.

Chapters 4, 5, 6, and 7 all relate to the Conjuration. Chapter 7 also relates to banishing.

Chapter 8 relates to the transcendental and cosmic implications of magic.

Magical Ritual part 1: Centring

The chattering mind provides the major obstacle to achieving magical intent. Only those who can learn how to switch off the internal dialogue can succeed with magic. If you can stop thinking you can do almost anything. (Includeing making terrible mistakes.)

Ordinary thinking proceeds dualistically, the moment a thought or desire arises it provokes its negation in the form of a doubt or a fear or a question or a further thought which modifies or distracts from the original.

Only unwavering intent will serve for magic, no matter if only achieved for a few minutes. For this we have to use parts of the mind other than the chattering parts which conduct the internal dialogue.

The centring manoeuvre consists of a move towards Gnosis, a technical term borrowed from first century western esotericism, and now used to mean that strangely abstracted state of single pointed awareness wherein virtually everything else in the universe disappears. Of course the concept also appears in eastern esotericism as well, where it features as the Dhyana of Raja Yoga and it has a variety of names including Samadhi in various Buddhist systems.

Perhaps we could describe it in Western terms as the achievement of a certain 'neurological quantum coherence' wherein some of the active parts of the brain enter into a single quantum state or something approaching it. In the context of what follows in Chapter 4 this may not seem as incredible as it sounds.

Some magicians can sometimes, with practise, simply close their eyes and centre themselves within seconds by a ferocious concentration on some simple visualisation or simply by concentrating on accessing a still point beneath the creative chaos within. However the ritual centring which follows provides a reliable egress into the centred mode at will.

It depends primarily on visualisation, breath control, and the vibration of sounds to temporarily paralyse the internal dialogue and prepare the mind for interaction with the microcosm and the macrocosm.

The magician visualises a sequence of images and vibrates certain sounds with unwavering concentration during the centring. The images have various attributions to metaphysical concepts and the sounds have various attributions to parts of the magician's symbolic 'subtle anatomy'. However these attributions should remain subconscious during the practise which only gives good results once learnt and practised till performance becomes perfect, unthinking, automatic, and untainted by wandering thoughts.

(Note that the subtle anatomy that corresponds to the varied and mutually contradictory oriental systems of chackras, meridians, chi energy, etc, has no objective reality in itself. You cannot locate it with a scalpel or a microscope. Yet it does provide a useful symbolic representation and therefore a means of manipulating objective reality magically. Magicians, doctors, and martial artists who wish to attempt more complex manoeuvres with the subtle anatomy often use more complicated models of it including extra chackras and meridians.)

The magician fiercely visualises each of the images in turn in the following chart beginning at the top whilst intoning the vibrated vowel sound that goes with it for a long out-breath. The magician should strive to sustain each vowel sound for an entire prolonged out-breath.

Having reached the bottom, the magician repeats that one and then works back up finishing with a final visualisation of the eight rayed star and a vibration of the Iiiiiiiii! 'mantra' to finish.

The following chart shows the images and associated vowel sounds. The sounds themselves tend to stimulate a sympathetic vibration in various parts of the magician's objective anatomy as well. Iiiii for the brow/'third eye', Eeeeee for the throat, Aaaaaa for the heart, Oooooo for the solar plexus, and a deep Uuuuuu for the peritoneal-genital area.

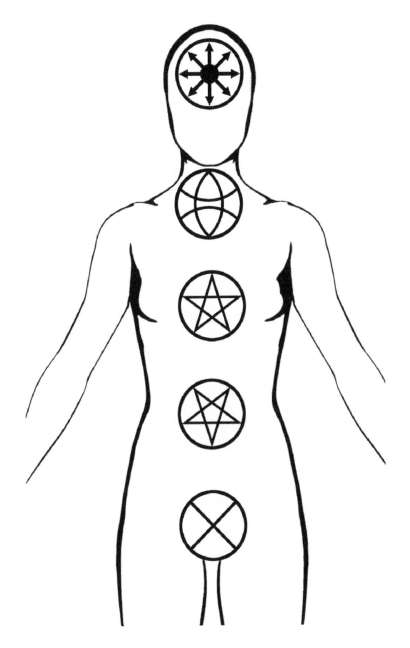

Fig 4. GCR diagrams

Note that in a variety of previously published rituals, some form of Centring gets repeated again after the Encircling. Exhaustive research indicates that this does not seem necessary and that it may even prove counterproductive if the magician intends to continue with further conjuration involving elements of invocation, evocation, enchantment, illumination or divination. The practise of Centring – Encircling-Centring seems to have derived from traditions where no further overtly magical activity immediately followed, and participants then went on to contemplate or meditate on other matters. However for direct follow through to further magical action, the balance of technical opinion now favours Centring, Encircling, and then straight to further conjuration, with the second centring reserved for a 'banishing' at the end if required.

The Magical Ritual part 2 commentary in Chapter 2 shows the procedures of Encirclement. The meanings of the symbols used in a chaos magic style Centring now follow. The magician should have a familiarity with these meanings but allow them to remain subconscious whilst actually performing the Centring.

The Chaos Star (Octaris) at the top represents the Panpsychosphere. Magicians, sorcerers and shamans tend to use the Panpsychic Hypothesis, the intuition that all phenomena behave in animate fashion to some degree, (depending on their quantum characteristics in the modern view). This; or something like it, constitutes the basic inductive leap from observation and experience that makes the magical paradigm.

The Panpsychosphere signifies all possibilities including apparent impossibilities, the realm of the aethers/astral planes (or quantum waves) where material reality branches into multidimensional Apophusis and where Apophasis occurs in which unreal events subtend real effects into the lower worlds. The Panpsychosphere represents the realm of cosmic imagination from which emergent phenomena arise spontaneously and chaotically in all the lower spheres and which stimulates personal creativity. Perhaps that barely makes sense in words, and the maths looks even worse, but it perhaps conveys the basic intuition to some extent. The panpsychosphere has a symbolic location at the brow in the position of a sort of psychic 'third eye'.

The Tellus, symbolically associated with the throat, represents the Noosphere. Sometimes called the Memesphere, it represents the sum of all ideas, beliefs, religions, philosophies, emotions, hopes and terrors arising from all structures capable of creating them. This primarily seems to arise from more complex organisms such as ourselves, but we should not discount the possibility that some of it may arise from other creatures with various degrees of intelligence. It symbolises all the information that exists in what one might call 'mental' form, thus its association with the throat, the organ of speech, where thought becomes communicable.

The Tellus symbol comes from an obscure astronomical glyph for the earth and it also suggests ideas converging, diverging, and overlapping.

The upright Pentagram represents the Anthrosphere, sometimes called the Anthroposphere or Opusphere. It represents humanity and our individual human-ness symbolised as a human figure spread-eagled as a five pointed star as in Leonardo Da Vinci's microcosm of Vitruvius. It also has the name Opusphere, the sphere of work, because it symbolises our most basic tool, our human form itself. The star lies centred on the heart area which supplies life sustaining blood to the five extremities of the human form.

The Biosphere resumes all living biological organisms from algae and bacteria to whales and mighty trees, including us. Basically this consists of a thin layer penetrating only a few tens of metres into the soil and a few hundred metres into the sky, although it has greater thickness in water.

The sophisticated image of Baphomet by Levi resumes plant and animal and human symbolism and hence we can use its simplified pentagramic form with two points upward here. The biosphere has obviously evolved from the geosphere and the Gaia hypothesis strongly suggests that the biosphere has modified to geosphere to meet its evolving needs. This star lies centred in the solar plexus area.

The Geosphere represents the apparently inanimate planet itself; this of course includes the lithosphere (rocks), hydrosphere (waters) and the atmosphere, plus the energy sphere, both in terms of the fiery core and the energy inflow and outflow with space. The symbol of a circle in a cross represents a traditional astrological sign for earth and also the quarters of earth, air, fire, and water comprising the planet. The magician locates this symbol in the genital-peritoneal area

Chapter 2
The Spells of the Spinning

$$W^2 = \frac{2\pi GM}{V_H} \text{ (Vorticitation)} \quad f = \frac{c}{2L} \quad \text{(Frequency)}$$

The hypersphere of the universe itself, and the myriad hyperespheres within it which constitute its fundamental particles, all share another defining characteristic, that of 'spin'. Whilst the spell of the binding holds the universe and its constituent particles together, the spell of the spinning prevents them from imploding into themselves. Thus everything on the macroscopic and microscopic levels exists in dynamic equilibrium between binding and spinning.

Most people will have a familiarity with how the orbital velocity of the planets prevents them from falling into the sun. Some people may have a familiarity with the idea that all fundamental particles have an intrinsic spin as well as a tendency to go into orbit about each other as in atoms. Few people can appreciate the idea that the universe itself also has a spin that balances its gravity that would otherwise cause it to contract with increasing speed into a cataclysmic implosion.

However the 'spins' of the microcosmic and macrocosmic realms differ somewhat from the readily observable spins of midicosmic phenomenon like toy balls or planets because hyperspheres spin about several axes simultaneously. We cannot easily visualise this but to differentiate it from the ordinary spin of spheres like balls or planets we can call it vorticitation. Unlike ordinary spherical spin that periodically transposes only some opposite surface points, vorticitation periodically transposes all opposite points, effectively converting objects into their mirror images and back again, as if through higher dimensions.

This vorticitation gives fundamental particles their properties of wavelength and frequency and it gives the universe the property of rotating back and forth between matter and anti-matter over a frequency period of about 11 billion years. (We cannot determine which state it actually occupies now but we designate it as 'matter', despite that the partial asymmetry between this matter and the 'anti-to-this-matter' that we can manufacture, indicates that we probably inhabit an epoch somewhere between the two extremes.)

The vorticitation of this universe acts rather like The Universal Clock in the Discworld universe. Whereas an ordinary clock merely 'tells what time it 'is', the Universal Clock 'tells time what it 'is'. In other words the vorticitation actually creates time. Our clock however comes with an unmarked dial as we inhabit a finite and unbounded spacetime, but we can place markers on it to suit ourselves.

The Vorticitation equation $W^2 = \dfrac{2\pi GM}{V_H}$ derives from an equation that Gödel [7] derived from the field equations of general relativity to show that a spherical universe could have a spin, but adapted here for a hypersphere rather than for a sphere.

In this equation W stands for the angular velocity in radians, G stands for the Gravitational Constant, M stands for the Mass of the universe, and V_H stands for the 3D hyperspherical 'surface' volume of the hypersphere.

The spell of the spinning gives a vorticitation rate of a mere 0.006 arc-seconds per century when we enter the universe's mass calculated from the Anderson acceleration, and hypersphere 3-volume, Vh. This 'spin' passes more or less unnoticed particularly as it does not have a visible axis of rotation. Yet it supplies sufficient centrifugal effect to exactly counter the centripetal effect of the small positive spacetime curvature.

When applied to fundamental particles the same spell shows the precise relationship between their frequency and wavelength: -

By substituting the hyperspherical 3-volume $2L^3/\pi$, and the Spell of the Binding equation,
$\dfrac{GM}{L} = c^2$ into $W^2 = \dfrac{2\pi GM}{V_H}$ and then substituting and
$W = 2\pi f$, to remove the radians, we obtain the corollary: f = c/2L, the frequency equation.

If we then equate antipode length with wavelength then this shows the familiar frequency f, lightspeed c, and wavelength w, relationship for matter (fermion) particles f = c/2w.

That fundamental particles run on the same equations as the universe itself does seem noteworthy to say the least. In fact it looks like an extraordinary vindication of old HermesT.

Thus - As Above, so Below, but what about the Midicosm?

Well the Macrocosmic vorticitation modifies the structures of galaxies and interferes a bit with our space-probe manoeuvres, but the main effect of this kind of 'rotation in all dimensions at once' lies in the effect that it has on time. It gives time 3 dimensions rather than just the one we usually acknowledge, but more of that in Chapter 6.

The microcosmic vorticitation possibilities lead to a rich zoo of fundamental particles, as not all of them vorticitate in all possible dimensions simultaneously.

This second part of the exegesis of The Map shows that any realistic map of reality must include its dynamic aspects. Everything spins, existence depends on spin, and nothing has any kind of static 'being'. Everything exists only because it spins, or because its components spin constantly. We inhabit a dynamic universe where nothing stays still.

This leads to the liberating and perhaps horrifying realisation that we do not have any sort of underlying 'being' now, let alone before or after death.

If you have ever watched an infant gradually acquire consciousness or an old person loose it to dementia, it becomes perfectly obvious that we do not contain any sort of soul or fixed inner self or 'being'.

As Heraclitus of Ephesus [8] noted, 'You cannot step in the same river twice', or as his student Cratylus reportedly once said, in a moment of Zen-like clarity and opacity, 'You cannot even step in the same river once'.

Hyperspherical vorticitation, the spin in multiple dimensions that maintains the universe and its particles, does not submit to visualisation very easily as hyperspheres lack perceptible edges and surfaces. (We cannot even readily perceive the spinning of an ordinary sphere unless we place some markers on its surface.) A spinning cube however readily makes itself apparent because we can see the edges and faces passing by. Thus a spinning hypercube can provide an impression of what hyperspherical vorticitation would look like.

See figure 4, The rotating hypercube. Note that we show it rotating through just one of 3 possible orthogonal planes. It can actually rotate through all 3 simultaneously in full vorticitation mode.

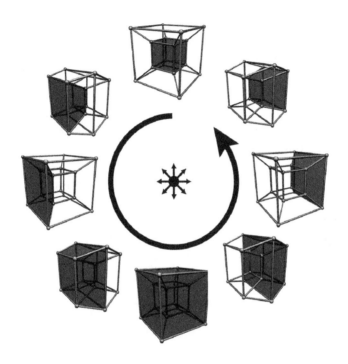

Fig 5. Rotating hypercube

A brief digression into the realm of spin now follows.

The conventional particle physics model of attractive and repulsive forces arising from the exchange of boson (force carrying particles) between fermions (matter particles) can perhaps explain repulsive forces but it does not provide a satisfying explanation of the mechanism of attraction. Repulsion seems easy enough to explain in terms of one matter particle emitting a force particle and recoiling as it does so, and then that force particle knocking another matter particle away when it hits it. However, how can such a mechanism possibly explain an attractive effect? How can one particle emit another which then has the effect of drawing towards it a third one?

The general relativistic model of gravity does however provide a satisfying mechanism for the attractive 'force' of gravity. This does not depend on matter particles emitting 'graviton' force carrying particles but rather on the idea that all particles consist of bits of curved spacetime. The celebrated rubber sheet analogy shows how this works. Two fairly heavy balls placed on a trampoline will create depressions in it and tend to roll towards each other. The trampoline represents space-time and the depressions correspond to the curvature in it created by the balls.

HD8 suggests that all the four fundamental forces actually act in a similar sort of way rather than by the mechanism suggested by particle physics.

The HD8 hypothesis suggests that all so called 'force' fields result from various types of spin in fundamental particles, but how could such fields; with both attractive and repulsive effects, arise from just spin?

Consider the classic rubber sheet model of gravity again. If instead of placing two weights upon a rubber sheet we get hold of two points on its surface and twist them then the tension between those two points will increase. Imagine boring a hole through the elastic sheet, attaching a small rigid disc on each side of the disc and then tightening a bolt between them so that they grip the sheet firmly. Alternatively imagine just thrusting your fingers between the fibres of a trampoline.

Then if you twist the bolt or your hand, it drags the fabric of the sheet around a bit. Amazingly it doesn't matter if you twist the fabric in the same direction or in opposed directions at various points, the tension between the points always increases. You can verify this easily enough by putting a couple of twists in an elastic band. This models gravity. Any point of spin or torsion in the fabric of spacetime creates some attraction for any other point of spin.

Now consider what happens if two twisted points in spacetime encounter each other. If the directions of the twists or torsions oppose each other then on meeting the twists will tend to attract and cancel each other out. In the extreme case where two particles with exactly opposite sets of spins meet they attract and annihilate each other.

If on the other hand the two particles have some spins that go in the same direction then they will tend to create an even bigger twist in spacetime when they approach, and spacetime resists this so they repel each other.

So if we can attribute charges like the electro-weak and the nuclear colour charge to various spins then like spins will repel and opposite spin will attract, despite that a small attraction exists between spins of any type that we recognise as gravity.

Thus we will not need to invoke virtual photons or virtual gluons as force carriers, spacetime itself will act as the medium.

Now as hyperspheres, fundamental particles have spins in various dimensions which we can call vorticitations. This means that they consist of points about which spacetime itself swirls with an effect that decreases with the inverse square of distance from the centre. Particles can have spins about several of the axes of spacetime and not all spins have the same dimensionality as others, so not all the particles in the zoo feel all the forces. Nevertheless the spin based 'force' model does provide an almost visualisable scheme to account for both attractive and repulsive effects.

We can perhaps imagine that spacetime consists of a sort of solid block of 'trampoline' material with various fibres running through it representing the dimensions of spacetime. Some of the spins appear as twists in some of the fibres but not others. Particles all respond weakly to the overall twist or torsion in the substrate (gravity) but more strongly, and in a directional sense, to the particular spins that they themselves possess. So all the chiral, electroweak, colour, and generational spins of particles contribute to the mass of such particles, but such spin-charges additionally have a particular interaction with neighbouring spins of the same dimensionality, either attractive or repulsive.

Thus matter tends to clump together under the effect of gravity until it reaches the point where repulsive forces between particles will prevent further collapse.

Thus as General Relativity replaced the Newtonian idea of a gravitational 'force' between all objects with the idea that all objects actually consist of bits of curved spacetime, so we can perhaps consider that all apparent fundamental 'forces' actually arise from spins of bits of spacetime which induce a torsion in the rest of it. However such a model does imply that we do need to use a map with eight dimensions rather than just the four of relativity. Such a map also appears to have the power to explain quantum superposition and entanglement and magical links. (See Chapter 5)

Such a model which explains all forces and particles in terms of the 'dynamic geometry' of spacetime has a history. The early Pythagorean mystery schools sought the numbers and shapes underlying reality. The Illuminati and the higher Freemasons hid their atheism behind the gloss of the Supreme Architect of the Universe, a concept which basically implies that the universe runs on potentially understandable geometric principles rather than on the incomprehensible whims of some cosmic sized anthropomorphic deity.

The Witches of the Ramtop Mountains on Discworld tend to disdain Wizardly theory in favour of herbalism and headology (psychology), however they do recognise that magical theory depends on Jommetry as they call it.

Even before Einstein formulated the equations of General Relativity, the formal idea of Geometrodynamics had received some attention. If he hadn't come up with Special Relativity first, and named it as such because of the significance of the relative positions and motions of observers that it showed, then he might well have named General Relativity as Geometrodynamics instead.

Some kind of eight dimensional spin mechanics would appear to offer a better prospect of unifying all the fundamental 'forces' in a single scheme than quantum gravity does. It could effectively heal the rift between the paradigms of general relativity and quantum physics and lead us on to the next phase of understanding; starships, teleportation, magic and all.

The Gift of Apophenia

Magical Ritual 2: Encircling

Apophenia's spacetime-dance.

Having Centred; the magician proceeds to the Encircling. In this the magician attempts to attune to the whole of the rest of the universe, to the microcosm above and to the macrocosm below, each of which stretches away from the midicosm of the centred will and perception.

Connecting with the entire universe remains theoretically improbable and practically impossible. Few magicians will have immediate concerns with events in distant galaxies or with events below the molecular scale. Nevertheless any magician can only interact magically with phenomena which exist, at least symbolically, in that magician's Map of Reality.

You can only have a functional magical link to something in reality if you have a fairly accurate representation of it inside your head.

Thus the Encircling really consists of an attempt by the magician to access the Map.

Historical practices such as the callings to the four quarters, or the drawing of magical circles, or circle dances, or spinning on the spot have become subject to various misunderstandings. They primarily serve to activate The Map in the magician's subconscious that forms a link to the universe outside. (Magical circles have little defensive value in our experience, only counter-spells really work).

The magician faces something of a dilemma when choosing suitable encircling techniques and symbolisms. On one hand it remains desirable to include as much of the known universe as possible, but on the other the ritual should remain fairly compact and easy to use. Magicians thus need to choose something which somehow resumes or précis their entire Map. The mere acknowledgement of geographical north, south, east, and west now seems inadequate. Drawing pentagrams to the four quarters may suffice for those who can attribute everything to earth, air, fire, water and aether, or in its more modern interpretation, to space, time, mass, energy, and information. The calling of various surrounding 'gods' and 'spirits' can also feature as part of a traditional encircling but this now seems better reserved for more precise work during the Empowering which follows after the Encircling.

Chaos Magicians tend to favour the use of the Octaris or Chaos-Star which for them represents all the major god-forms, plus the elements, plus the full dimensionality of the universe and its indeterminacy, whilst retaining a relatively simple shape for visualisation.

Such simple glyphs of the entire map work by subconscious Apophenia. The magician relies on them to connect with The Map stored in the subconscious, rather as the cover of a well read book serves as a reminder of its contents, or as an algebraic symbol such as 'a' for acceleration, can resume an entire concept.

Thus in a typical Chaos Magic style Encircling the magician may turn anticlockwise as we inhabit a universe where anticlockwise spins mainly predominate at all levels during this current phase of the vortivcitation cycle. See 'The Left hand of Creation'[9]. The magician draws large visualised chaos stars to the quarters and perhaps then above and below for good measure also. It helps considerably to perform this operation with closed eyes unless you have the rare ability to visualise figures hanging in the air without becoming distracted by the actual scenery. Some people can do this whilst looking against a blank surface or the sky.

A pointed finger or an object designated as a magical instrument helps, as does reinforcing the visualisation with some kind of vibration or mantra. In the case of the Chaos Star the IEAOU vibration again serves well, with one vowel intoned for each of the four intersecting lines and with the final U delivered to the centre.

Chapter 3
The Spells of Illusion

$$z = \frac{\lambda_o}{\lambda_e} - 1 = \frac{GM}{c^2} \frac{1}{L-a} - 1 \quad \text{(Redshift.)}$$

$$L = 1 + \sqrt{d - d^2} - d \quad \text{(Hyperspherical lensing)}$$

Since at least the days of the Classical Greeks and the early Vedic and Buddhist thinkers, metaphysicians and magicians have suspected that we do not perceive and model reality very accurately at all, and that we labour under manifold illusions. The simplest of tricks became metaphors for the idea that we may have very poor maps of reality indeed. Sticks dipped in water appear to bend where they cross its surface and shadows can easily deceive.

More recently some have speculated that this entire universe may arise as a holograph projected from higher dimensions or even consist of a simulation in some information processing system, run by various benign or malignant agencies.

Frequently the 'argument by analogy to illusion' evolved into an argument for the existence of various metaphysical phenomena such as gods and higher realms, and even life after death. However it has never proved possible to eliminate anomalies and inconsistencies from such speculations except by relegating or elevating them to the status of 'mysteries'.
Yet only by investigating anomalies and inconsistencies in our illusory maps and world-views can we gradually claw our way towards more accurate ones. Mysteries should present challenges, not opportunities for dumb belief.

This chapter examines illusions which arise from the structure of the universe itself and illusions which arise from the structure of the human mind. These can work together to create even greater illusions such as the famous 'God Delusion' [10].

A universe which has finite and unbounded extent in both space and time does not require an almighty omnipotent and omniscient creator deity.

A mind which consists of various modules monitoring each other does not require an immortal soul.

On the other hand the 'theory of mind' circuits in the human brain do have their uses, and we can put them to more modest and effective uses by making our own magical gods. This seems preferable to squandering them on the megalomaniac delusion that some non-existent creator of this stupefyingly vast universe deals with us personally, dispensing life, death, fortune, earthquake and disaster at whim.

Firstly to the question of illusions created by the structure of the universe:

The first equation deals with the illusion of an expanding universe. This still dominant hypothesis has formed the basis of cosmological thought for well over half a century now. It requires that the universe exploded out of some *in principle* unknowable cause to its present size from some *in principle* unknowable initial condition in a so-called big bang.

The observed cosmological redshiftng of light usually gets cited as prime evidence of the expansion of the universe. Light from distant galaxies appears to have downshifted in wavelength towards the lower energy red end of the spectrum by the time it gets here. As the redshift appears roughly proportional to distance travelled and because they think that distance itself shouldn't create the effect on its own, cosmologists have surmised that the underlying fabric of spacetime must have expanded to stretch out the light waves and thus downshift their wavelengths.

However the first of the illusion equations explains the redshift more elegantly,

$$z = \frac{\lambda_o}{\lambda_e} - 1 = \frac{GM}{c^2} \frac{1}{L-a} - 1 \quad \text{(Redshift.)}$$

Here the redshift z which means $\frac{\lambda_o}{\lambda_e} - 1$, the ratio of the observed wavelength over the expected wavelength, (minus one, so that the redshift scale starts at 0 rather than 1), equals the gravitational constant G, times the mass of the universe M, divided by lightspeed squared, times the reciprocal of the hyperspherical antipode distance L, minus the astronomical distance in question a.

Thus it expresses the redshift as a function of distance and provides a mechanism to explain it. In this model the gravity of the universe at large causes the redshifting of light passing across it. This should not surprise us, we already know that light becomes slightly redshifted when climbing out of a star, and that the degree of redshifting depends on the mass of the star.

The second illusion equation explains the apparent mismatch between redshift values and distance values estimated from magnitude at vast distances. Basically, very distant objects in the universe appear fainter than they should. Cosmologists have interpreted this as an acceleration in the expansion rate of the universe and they have invoked some form of mysterious phlogiston type stuff called 'dark energy' to explain it.

Yet if we inhabit a hyperspherical universe then its small positive spacetime curvature will create just such a mismatch at large distances. This arises because light itself has to follow the curvature of the universe and we usually operate under the assumption that light travels in straight lines.

Thus in a way this begins to look like a case of a stick dipped in water writ large. Spear-fishermen long ago learned to correct for the distorted apparent position of a fish under the surface and have thrown their spears short at an angle-corrected distance ever since. Unfortunately the majority of cosmologists have yet to correct for the curvature of the universe which distorts our view of it as shown in Figure 6.

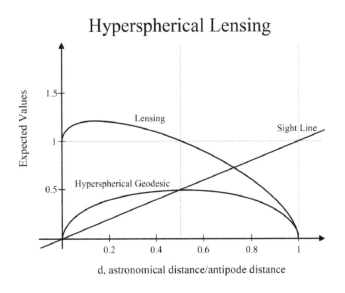

Figure 6. Hyperspherical Lensing

In this figure the origin represents the position of an observer looking out into space to the very extremities of the universe. The observer assumes that all incoming light has come in straight lines, thus the diagonal line represents the observer's expected sight line.

The bow shaped line represents the curvature of a hyperspherical universe, and light must actually have come to the observer along such a line. Now the asymmetric curved line represents the difference between what the observer expects and what the observer gets, it represents the amount of lensing that the giant hypersphere of the universe imposes, and this depends on distance.

The lensing has negligible effects on objects within millions of light years of the observer, so it does not affect the appearance of nearby galaxies. Observers looking at light coming from about a quarter of the way across the universe will actually see galaxies from a wider visual arc than they expect compressed into their visual arc.. Thus galaxies at these distances will appear slightly brighter for their actual distance by about 20%. Light from galaxies about halfway across the universe comes to an observer with no lensing and galaxies there will have the correct apparent magnitude for their distance. However light which comes from beyond the halfway point becomes subject to a progressively larger distortion because the diminishing visual arc gets spread out over an ever-widening field of the observer's view.

Thus objects past the halfway distance to the other side of the observer's universe will have a much fainter magnitude than they should for their distances as estimated from their redshifts.

$$L = 1 + \sqrt{d - d^2} - d \qquad \text{(Hyperspherical lensing, algebraic form)}$$

L means the hyperspherical lensing at any particular distance,

d means the fraction $\dfrac{a}{L}$, the astronomical distance a, over the distance to the antipode, at about 11 billion light years.

Orthodox cosmologists have yet to appreciate this vast optical illusion and have instead opted to resolve the anomaly by hypothesising that the universe has entered an accelerating phase of expansion and must therefore contain a stupendous amount of otherwise undetectable energy (dark energy) to drive it.
The hyperspherical model of the universe at least has the virtue of simplicity. If it has a finite and unbounded extent in space and time then the universe doesn't require a beginning or a creation (or a creator), or some otherwise undetectable sort of dark energy.

In this universe very little actually exists in the form in which we perceive it, the fish-eye lensing effect of the hyperspherical curvature of space merely provides one of the most spectacular illusions, but time provides perhaps the most peculiar illusion of all, as chapter 6 will attempt to show.

Occultism and Illuminism can have variable meanings. In one sense all the best scientists and magicians practise occultism, but neither relish the term 'Occultist' because of its association with mystery mongering and cultism. Yet they investigate the unknown, the hidden, and that which seems occluded, in search of illumination.

Seek those in search of Illumination, but shun those who claim to have it. Nobody can claim Illumination with so many mysteries left unresolved or simply smothered in further imaginary mysteries.

Much of what passes for Illuminism seeks not to unravel real mysteries but merely to impose belief structures upon existing ignorance. All religions have ignorance masquerading as mystery at their cores.

On a whimsical note we should remember that the inhabitants of Discworld suffer every bit as much from various illusions as we do here on Roundworld; however they have a specific antidote for them, Kalachian Coffee. This stuff makes you Knerd, or anti-drunk. Thus on Discworld the spectrum of intoxication runs from Drunk to Sober to Knerd. Thus sobriety just means living under a few less of the comforting rosy illusions that come with drunkenness. When someone becomes knerd, all illusions disappear to reveal reality in the raw and the victim often screams continuously till the effects wear off.

(Apparently our Roundworld synonym 'Nerd' devolves from Silicon Valley slang for those who don't, or who shouldn't do drink or drugs at work.)

It would seem that a confrontation with the prospect of cowering on a tiny planet with a fragile ecology spinning in a radiation blasted void with no omnipotent benign gods, no immortal soul, no real self, and an inevitable death, proves rather too much for most people.

In seeking a cure for religion we need to look at what causes it and what makes it continue to function. Religions attempt to answer the question 'Why'. The question 'Why' invites a meaningful story for an answer rather than a mechanistic explanation. In the absence of mechanistic explanations people generally prefer stories to no answer at all.

By asking 'How' science attempts to develop mechanistic explanations and we now have mechanistic explanations of storms, earthquakes, and even the complexity of life on this planet, which don't involve stories about gods at all.

However neither Why-type answers nor How-type answers fully satisfy any question. Why-type answers always end with 'because Amon-Ra, Odin, or Jehovah (or whomsoever) wills it!' How-type answers similarly tail off through causal chains ending either with quantum indeterminacy or with the inexplicable initial conditions of a big bang.

Magic asks another kind of question; the question 'What'. This phenomenological line of enquiry forks into two complementary streams, the 'Essentialist', 'what does this phenomenon resemble' (the Doctrine of Signatures or Apophenic mode) and the 'Existentialist', 'what does this phenomenon actually do' (the purely Phenomenological mode).

This magical style of enquiry provides a new perspective on gods: -

Essentialist - Gods bear a very suspicious resemblance to stuff in our own heads.

Existentialist - The power of gods depends entirely on our willingness to believe in them.

Pratchett makes the latter point time and time again about the Discworld's gods; one might almost suspect him of blasphemously teasing us about our own.

[Note - Only the magical paradigm can really accommodate randomness. You cannot *in principle* explain 'How' randomness occurs nor 'Why' because randomness implies the absence of mechanism or meaning. Chaos Magicians on the other hand generally feel quite comfortable with randomness. 'What does this do?' 'Aha it acts randomly, well fine, I often act randomly myself!']

So, in the realisation that the universe does not require a creator, most magicians have discarded the Monotheist Phallacy of some big-daddy Uber-God who lurches wildly through benign-sentimental and megalomanic-genocidal mood swings. There seems little point in having a god that doesn't often do what you want, but merely represents what the tribal elders want of you, plus a generous helping of their own ignorance.

The very existence of the universe or indeed of anything at all, has often served as justification for the existence of gods, although the ancient Greeks mostly seem to have modestly avoided claiming that their gods actually created The Entire Universe.

Yet carefully constructed gods do provide useful resources of inspiration and power for those who invest belief in them. Magicians create gods using programs they find in their own heads and often from bits of various mythologies they find lying around as well.

The illusion of gods has proved persistent throughout recorded human history because humans have evolved a brain that contains a rather useful 'theory of mind' function. It can attribute 'intentionality' to other humans and to animals and perhaps to other phenomena, and thus it develops strategies for anticipating their actions, intentions, desires, lies and deceits. In other words, it has the ability to conceive of various phenomena as possessed of 'minds' which may have agendas other than that overtly betrayed by their immediate behaviour.

Rather curiously, humans learn to apply theory of mind to others before they apply it to themselves, and so ideas corresponding to self-awareness tend to develop only after the attribution of 'mind' to others. (Consequently we tend to develop self-images based on our constructions of the 'minds' that we attribute to our peers).

Of course the 'minds' that we attribute to other people always remain something of an abstraction and an approximation, we can never experience their thinking directly. However and here comes a really weird bit, we never 'directly' experience ourselves thinking either, self-awareness always consists of one part of the brain noticing what another part does. So when we turn our theory of mind facility upon ourselves we merely create yet another construct, either the desired singular self-image of the monotheist or the more realistic multi-self image of the neo-pantheist or magician.

Most pagan and polytheist pantheons supply a range of deities that resume most of human psychology and often deities that resume the strange powers of various orders of animals as well.

The Gods and Titans come from Chaos as the ancient Greeks seem to have understood. They arise in our minds as analogues of the random creative processes which underlie the quantum realm, natural phenomena, biological and psychological evolution, and ultimately cultural evolution. Successful invocation of many of these god-forms can bestow upon the magician rare insights and on occasion even grant access to unusual abilities that have evolved in our own subconscious minds.

The universe appears to owe its existence to the duality between binding and spinning as suggested in the first two chapters, rather than to some singular phenomenon like a Big Bang origin of reality.

One might argue that everything in manifestation only phenomenises because it lies suspended somewhere between opposing forces.

Duality seems to lie at the root of anything that we can perceive, nothing seems to exist except in the context of its opposite or its complementary quality.

As complex organisms ourselves we arise defined by sex and death, the one creates us and the other removes us. Both have led to the evolution of our complexity, and both seem to supply so much of our motivation, our desires and fears, and our internal dialogue. It also supplies most of our entertainment, if sometimes in sublimated form, because most of us do not like to admit our fascination with direct sex and slaughter in polite company.

Whenever you encounter a human culture extolling some moral virtue you can expect to see its relative absence actually characterising most of the everyday behaviour.

Compassion only gets valued where casual brutality and self-interest remain the norm.

(Compassion had extreme personal survival costs and only minor collective benefits on the harsh Tibetan plateau before the Chinese industrialised it).

Remain wary of cultures that extol Peace, they usually have the most weapons.

Sentimentality remains a sure indication of an enhanced capacity for cruelty.

(Just look at the nauseating folksy atavism and all the sentimental songs that seem to precede any aggressive military adventure.)

Patriotism consists at least of as much fear and hatred of other people's cultures and beliefs as it does of love of ones own. (Any New Reich will need enemies as much as it will need friends, as Josef Gobbels [11] observed).

Our perception of our personal self arises from a whole suite of emotional identifications, but as highly suggestible creatures we remain open to persuasion, and we don't behave particularly consistently. However we do tend to enter into temporary states of self-identification with whatever we do find ourselves doing.

All but the most autistic of us has, through socio-biological necessity, a 'theory of mind' capacity which allows us to make mental models of other people so that we can anticipate their likely behaviour and to some extent their feelings and their lies.

All but the most socially adept have fairly simple models of other people. Actors, politicians, and celebrities can fool us all too easily.

However we do tend to have rather more sophisticated models of our own selves, and note that 'selves' appears here as plural.

There seems no evidence that we contain some sort of 'master-self theory of mind department' which decides, okay so one part wants that, another wants something different, and yet a third has another desire, so I Will Take A Decision.

Rather it seems that our theory of mind facility does not have some sort of infinitely recursive theory of itself; but rather that it acts to provide each of the models with a model of the others. So we seem to consist not of a dictatorship but more of a parliament with its own built in opposition, apart from the most bloody-minded of us of course.

Thus from a general consideration of universal dualistic principles and a consideration of ourselves as fundamentally homeostatic organisms suspended between various sets of non-survivable conditions we arrive at: -

Stokastikos' Law
'All humans (and probably all sentient structures in the universe) have an Even number of selves'

This may not appear immediately credible but consider the following: -

Even the simplest responsive system in the information technology or cybernetic sense has to have two partly independent subsystems, and some sort of feedback mechanism between them. (See thermostats or the theory behind computer design.)

You don't really know anyone at all until you know something about their positive and their negative preferences, and what they might do under non-ordinary circumstances.

The supposedly omnipotent and omniscient monotheist gods make no sense whatsoever unless you relegate the problem of evil to that of incomprehensible mystery. You really need a Devil in the theory as well, but then you get the additional problem of the duality between 'fun' evil and the nasty stuff that hurts other people. Monotheists cannot really avoid sympathy for the Devil, it comes with the whole sin/grace territory, but it blinds them to the difference between fairly harmless fun and intellectual curiosity, and crusade, jihad, genocide, child abuse, and the stoning or burning of people at the stake.

Personally I don't trust monotheists to act with any sense of the full breadth of their humanity, they often seem monomaniac and deficient in humanistic terms, particularly when they have, or have had, secular power.

On the other hand militant atheists don't attract me either. Whilst they do a good job of trashing the monotheists and their imaginary 'real' deities they frequently fail to offer anything beyond their simplistic and supposedly almost closed materialist paradigm. Heck no, surely we haven't got this far just to enjoy material comfort. A vast universe awaits us full of all kinds of weirdness, magic, and opportunity, and we have barely fumbled the keys into the front door yet.

As humans we construct theory of mind illusions about other people, ourselves or ourselfs, and other phenomena ranging from animals to natural phenomena that some would regard as inanimate, and perhaps even to a theory of mind for everything in reality to make a fully developed panpsychism. 'Theory of mind' projections allow their users to enter into a more understandable or user-friendly relationship with the phenomenon concerned, rather than to remain somewhat autistically detached from them.

To access such powers magicians need to explore their own Psychotheology and then to act as Priests and Avatars of their chosen gods. It may appear quite deranged to civilians that anyone might choose to create a few Gods or Goddesses out of their own impulses and bits and pieces of mythology, but in fact humans have always done this, how else do you think they do it? Divine revelation just means getting really excited when you do it, and we can arrange that technical matter fairly easily. (See 'Gnosis' in chapter 7)

Religious types invariably try to convince others of the reality of their own gods because they want followers. Magicians usually keep them to themselves, (followers usually prove irritating) but they may pass on useful tips to friends and colleagues, so we'll present our favourites briefly in the order in which we got to know them and with paragraphs whose lengths indicate their relative importance to us.

Baphomet takes its name and image and some of its mythos from the 19th century Mage Eliphas Levi[12] who drew on various inspirations including the myths surrounding the Knights Templar. The androgynous theriomorphic goat-human image has also found favour with some contemporary Satanists. We prefer to invoke it as embodying our animal vitality and sexuality and the accumulated wisdom of biological evolution. An Invocation to Baphomet appears in the Mass of Chaos (B), see *Liber Null & Psychonaut*.

Eris, appears as a minor Greek goddess, a sister to Ares the god of war, a sort of female version of Loki given to making mischief and initiating upheaval and renewal. She recently got elevated to major goddess status by chaoist psychotheologians in the already legendary *Principia Discordia*[13]. An Invocation to Eris appears as The Mass of Chaos (E) in the Appendix.

Ouranos personifies our magician identity as opposed to our affable everyday solar self.
He takes the planetary symbol of Uranus as his sign; he lurks beyond the realm of the seven classical planets as an outsider with his spin orientated orthogonally to those of the rest of the system, like his planetary namesake. A shadowy hooded figure, he orbits in the outer darkness researching forbidden knowledge, hidden knowledge, and antinomian wisdom, the strange stuff that regular people

despise, fear, or ignore. An Invocation to Ouranos appears on the Ouranos audio CD [14].

Azathoth, appears as The Blind Mad God at the Centre of Chaos from the Cthulhu Mythos [15]. Invoke it only in dire emergency when all else has failed and only an unpredictable and extreme event may save the day. An Invocation to Azathoth appears in *Liber Kaos.*

Apophenia takes her name from a psychological quirk that many magicians have or cultivate, the ability to perceive meanings and connections between various phenomena where others do not, and as such it underlies both psychosis and genius. We invoke her in the form of a beautiful muse and she has responded with numbers and symbols and further names by which we may call her for inspiration. Her Invocation appears in *The Apophenion.*

Higher Chaos, the stuff that bears the Ouranian-barbaric name of OBDAXAZONGAGA.

Whilst the dreaded Azathoth represents the madly random and frequently destructive manifestation of chaos, the OBDAX represents the creative and complexifying stochastic manifestation of Chaos, and a work in progress for us, see Appendix 1.

This grimoire comes forth in large part by the grace of Apophenia and Ouranos.

We invoke Eris only by supplication, the others we invoke by personification or full possession. Baphomet and Ouranos each have dedicated servitors with groundsleves for the execution of relevant enchantments and divinations. I hope to avoid getting into situations which may necessitate any further Azathoth conjurations.

(Groundsleves consist of physical representations of entities, normally in the form of smallish figurines.)

To function as a Priest or Avatar of one of these syncretic or refurbished or discovered god forms or *Chaomeras* as we sometimes call them, takes a considerable amount of work, imagination, and devotion, yet it well repays the effort if conducted skilfully. Magicians should seek Chaomeras which resonate with parts of the pantheon of their own psychology and with their life aims, and learn, or devise their own invocations to them. It often helps to make symbols and representations of them in sculptures or paintings, or in other media. Such idolatry serves a useful purpose. It serves to link the curious hidden powers and abilities which lurk in the subconscious to a series of anthropomorphic representations with which, and through which, the magician can more readily interact.

Magicians remain unconcerned with the question of the real or illusory nature of their gods. The real or illusory nature of gods matters not, it matters only what it resembles and what it does, what inspirations and powers and abilities it brings. The more theoretically minded sorcerer-scientists may well ask how it does these things. Opinion here has swung almost entirely away from the independent gaseous vertebrate 'spirit' model of the religious paradigm and largely away from most species of the 'astral resonance' or 'collective unconscious' ideas that accompanied the scientific paradigm towards an entirely personal psychology model with a bit of parapsychology thrown in to explain the weirdness. In other words most magicians explain the doings of their gods as miracles of faith - their faith in them.

Magical Ritual part 3

So in a full magical ritual, having Centred and Encircled, the magician proceeds to The Empowerment. The magician chooses an Empowerment relevant to the intended Conjuration unless the ritual has no Conjuration.

A Conjuration-free ritual lacks any formal intent to follow through with an enchantment, a divination, an invocation, an evocation, or an illumination.

Magicians often perform conjuration-free rituals as daily practise when they don't have anything specific that needs doing. Instead they perform the ritual as a magical workout that also acts to empower their gods by giving them periodic attention (~worship).

Thus in conjuration-free ritual the magician will usually call on several god forms, for example by turning towards various segments of an imagined surrounding circle or sphere and calling one to each. Few magicians attempt to maintain relations with more than eight. Three, four or five will usually suffice.

During the Empowerment the magician calls the gods by their appropriate names and symbolic signs and numbers and in their visualised form whilst uttering or mentally reciting any appropriate litany or invocatory preamble.

In a full magical ritual which includes a Conjuration the magician will normally call upon a single god form whose powers and knowledge fit the intended purpose of the Conjuration as closely as possible.

The magician can attempt any degree of empowerment ranging from a mild summoning of a chosen power source to a full invocation with outright possession in which the magician loses consciousness of anything else for a time.

The process begins with Mythogenesis and proceeds to Summoning and can end with Invocation.

In mythogenesis the magician explores and develops the story of the god form or thought form or *Chaomera* as we sometimes call it, fleshing out its mythos and thinking about its symbols and powers and attributes. Classical mythologies provide plenty of source material, as do other more modern sources of fiction; or the magician may decide to create something afresh. In all cases magicians need to find something which has at least some resonance with something within themselves which they wish to amplify.

Summoning proceeds by some kind of a call to the god form by the use of visualisation, imagination and incantation, which the magician may supplement with the drawing or exhibiting or visualisation of symbols and instruments and the use of particular incenses or perfumes.

In a full invocation the magician proceeds to actually embody or incarnate or personify the god form as completely as possible. This necessarily involves a certain craziness and a temporary suspension of disbelief and ordinary identity. It involves the magician making a supreme effort of method acting and attempting to become possessed by the invoked form. 'Fake it till you make it' as we say in the trade; and you just might surprise yourselfs with what inner resources they have.

The Magical Ritual sections of the remaining five chapters all deal with various aspects of Chaometry, the art and science of making conjurations work effectively. Till now we have just looked at the preparatory part of magical ritual that cannot really go wrong. Conjuration sometimes fails to give any result at all, and sometimes it gives disastrous results. The Chaometry sections will examine the theory and

practise of improving the magician's ability to conjure for
desired results.

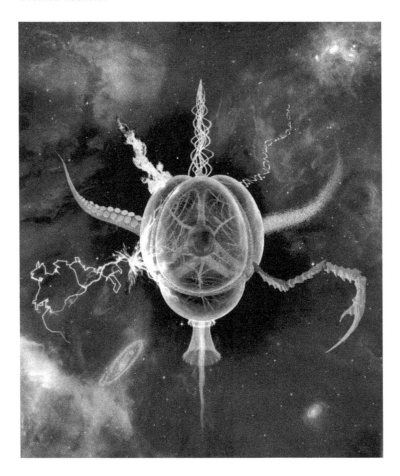

Azathoth

Chapter 4
The Spells of Subtle Magic

$$\Delta E \Delta t \sim \hbar \sqrt[3]{U} \qquad \text{(Energy)}$$

$$\Delta S K^0 \Delta t \sim \hbar \sqrt[3]{U} \qquad \text{(Entropy)}$$

$$\Delta H k K^0 \Delta t \sim \hbar \sqrt[3]{U} \qquad \text{(Information)}$$

Discworld has a very low value for **c**, lightspeed. Light there only moves at about 200mph, whereas on our Roundworld and in the surrounding Hyperspherical universe it maintains a rather more impressive 186,000 miles per second in a vacuum, and only a bit less in air. Magic works easily on Discworld but causality doesn't work all that reliably. Here on Roundworld we have it the other way round, causality works reasonably well but we really have to struggle to get magic to work.

Now a low value for Discworld's lightspeed implies that it should have a much higher value for Planck's constant $\hbar$, because: -

Either $\hbar$ times c has to remain fairly constant or the two worlds wouldn't resemble each other faintly, even in fiction.

Or perhaps because as **c** roughly defines the amount of Causality in a universe and $\hbar$ roughly defines the amount of Chaos, then if one goes down the other has to go up.

Either way, Planck's constant, which marks 'quantum' grain-size of the universe; indicates the sort of level where causality breaks down and chance and magic take over.

Now Roundworld scientists have often argued that because Planck's constant has such a small value here, we can safely ignore quantum effects in the human brain or indeed on the midicosmic scale of human sized events generally.

However it now looks that the effective grain size of our universe actually kicks in at a considerably higher level than that suggested by Planck's constant and that we inhabit a universe with much bigger pixels because the universe cannot contain enough information to make smaller pixels behave causally.

Consider the following: -

$$S = -k \sum_i p_i \log p_i \quad \text{Boltzmann-Gibbs}^{[16]}$$

$$H = -\sum_i p_i \log p_i \quad \text{Shannon}^{[17]}$$

The Boltzmann-Gibbs equation states that Entropy, S, of a system relates to Boltzmann's constant k, (which converts between energy and heat) and a nasty looking integral which defines the probabilities of the microstates of the system.

The Shannon equation states that the Information, H, in a system equals the same thing but without the k.

This leads to the conjecture that H = S/k and that a deep relationship exists between Entropy and Information. Some theorists think in terms of Information as Negentropy.

However whereas the entropy equation usually refers to the molecular microstates of a system, the microstates of an information system may relate to anything from the choice of letters in an alphabet string, to a string of coin toss results. Information often only seems to exist in the context of something able to interpret it as such. Chapter 8 deals with this point in more detail.

Nevertheless the smallest units of reality down at the quantum level presumably represent the smallest possible units of information storage even in the absence of anything able to interpret the information as such.

Beckenstein and Hawking conjectured that black holes could not destroy information when sucking things in, but because sucking things in would increase their radius the entropy of a black hole would have to depend on its surface area in Planck units: -

$$S_{BH} = \frac{kA}{4l_p^2} \quad \text{Beckenstein-Hawking}^{18}$$

Here the entropy of a black hole S_{BH} equals Boltzmanns constant times the surface area of the hole, A, divided by the Planck area times four.

Now we can more or less forget about the four in an order of magnitude calculation because of the gigantic size of $\frac{A}{l_p^2}$ and if we accept that $H = S/k$ then the equation reduces to:

$H = \frac{A}{l_p^2}$ where the information contained within a black hole equals its surface area in Planck units.

This intriguing idea leads to the Holographic Universe Conjecture which states that the amount of information in a universe relates to its surface area. Now this conjecture fits the expanding universe hypothesis rather uneasily because the concept of a surface of such a thing would merely correspond to its expanding visible horizon.

However a hyperspherical universe has a well defined and easily quantified surface, whether we consider 'surface' to mean its 3 dimensional hypersurface or the more ordinary sort of surface that a hypothetical observer on its outside might see. The ratio of information comes out the same whether we calculate it in terms of bits per hypervolume for a hypersurface or bits per volume for a surface, as follows: -

As the Antipode length of the universe equals the Ubiquity constant U, times the Planck length then $\dfrac{A}{l_p^2} = U^2 = H$, so the information content of the universe comes out at a whopping 10^{120} bits. However this has to suffice for all the 10^{180} Planck volumes, (ubiquity cubed), within it, so that means only about one bit per 10^{60} Planck volumes. Thus the universe has a serious information deficit; it does not contain enough information to causally specify events below a certain grain size which corresponds to about 10^{20} Planck lengths, i.e. the cube root of Ubiquity, $\sqrt[3]{U}$ Planck lengths.

Now this actually accords rather well with observation, we do not actually observe any real quantities below these levels anyway. Although fundamental particles theoretically exist as dimensionless points they have effective sizes down at the 10^{-15} metres size when they interact, and similarly nothing ever seems to happen in less than 10^{-24} seconds. These lower limits seem to represent the actual pixilation level of the universe, its effective 'grain size', and any real event either occupies a whole pixel or it doesn't happen. This minimum level of divisibility represents the point where causality ceases to apply and processes become indeterminate. Thus it acts a whole lot more chaotically than Heisenberg thought.

For many decades it has seemed from theoretical considerations that true Chaos or indeterminacy applies only at the Planck length scale of $\sim 10^{-35}$ metres or the Planck time scale of $\sim 10^{-44}$ seconds. This hardly accords with either life as we experience it or what we know about the behaviour of fundamental particles or our own minds. It actually seems to kick in at a level twenty orders of magnitude ($\sqrt[3]{U}$) higher.

Thus for practical purposes we can factor in those twenty orders of magnitude into some of the Heisenberg[19] Uncertainty/Indeterminacy relationships by using the enhanced value for Planck's constant, $\hbar \sqrt[3]{U}$.

$$\Delta E \Delta t \sim \hbar \sqrt[3]{U}$$
$$\Delta S K^0 \Delta t \sim \hbar \sqrt[3]{U}$$
$$\Delta H k K^0 \Delta t \sim \hbar \sqrt[3]{U}$$

In the first of the subtle magic equations the indeterminacy of the energy, E, times the indeterminacy of the time, t, yields about 10^{-14} joules available for about 1 second. That may not sound like a lot, but typical nerve conduction energy equates to $\sim 10^{-15}$ joules and a typical nerve transmission time between neurones takes $\sim 10^{-3}$ seconds. Thus it provides enough to power 10,0000 neurones, quite enough for a substantial thought to appear spontaneously every second.

At this sort of level we no longer have a problem finding quantum activity in the brain but rather we have the problem of explaining why it doesn't overwhelm us. Well the brain certainly seems to have damping mechanisms in the chemical synapses between the neurones, as revealed by sufficient hallucinogens to turn them off, and then you do often seem to get substantial thoughts appearing out of nowhere every second, more or less uncontrollably. Magicians however have more interest in achieving a more selective use of this function when seeking personal visions or inspirations or implanting thoughts in the minds of others.

For reasons discussed in the whimsical addendum to this chapter we can take the liberty of identifying the 10^{-14} joules of energy per second as a basic magical energy unit, the Prime, denoted by the Greek uppercase letter Π, (Pi).

The second of the subtle magic equations arises as a corollary to the first and it relates to the indeterminacy of entropy, S, with time, t. To balance the dimensions of the equation we need to multiply the entropy by the absolute temperature K^0 taken as 300^0 Kelvin, (27^0 Celsius). The indeterminacy appears as either plus or minus to allows for entropy decreases as well as increases in either direction in time.

Now just how much entropy reversal we can get out of this remains rather difficult to quantify in terms of bits of information because it depends on what the 10^{-14} joules per second can do to the microstates of the system in question. In the brain it can do quite lot, but in the lumpier world outside it can only affect finely balanced events.

The third of the subtle magic equations relates the indeterminacy of information and time to the enhanced Planck constant .by substituting Hk for S, and it yields an astonishing result, $\Delta HkK^0 \sim 10^7$ bits per second. However this represents the maximum possible information indeterminacy of a system where the grain size and transition energies have exceedingly low values. The large value of the information results from the very small value of

Boltzmann's constant which relates to the molecular level, and the magician will rarely want to interact with that, you cannot usually even see a mere 10^7 molecules. In practise the magician needs to factor in a considerably larger number corresponding to the number of useful states of a system and the transition energies between them, to find out how much information in the system remains open to psychic interference.

We can regard the 10^7 bits per second as a basic unit of magical information transfer and denote it as the Thaum, symbolised by the Greek uppercase letter Θ , (Theta)

Thus we arrive at the sobering conclusion that although this universe contains enough Chaos to allow magic it doesn't contain enough to permit gross miracles in a hurry.
The magician will need to target events which depend on very small energy or entropy changes and the results won't often look much like spectacular parapsychology, they will look more like a series of events going somewhat improbably in the desired direction.

Baphomet

Magical Ritual Part 4: Chaometry (i)

When selecting the objective for the Conjuration phase of the ritual the magician needs to enchant or divine for events or knowledge that require only small energy or information or entropy effects. For example the spontaneous materialisation of gold requires prohibitive amounts of energy and even the apport of gold usually requires a considerable alteration to the information structure of the universe. On the other hand conjuring a business worth a lot of gold bars out of virtually nothing merely involves a change of attitude and changing the thoughts of those you want to involve in the enterprise as suppliers or customers.

Similarly in divining the unknown the magician needs to reduce the uncertainty to manageable levels, primarily by finding out everything possible about the situation first.

As the tried and tested maxim goes, Enchant Long and Divine Short.

In Enchantment you have 10^{-14} joules per second of indeterminacy to play WITH.

In Divination you have 10^{-14} joules per second of indeterminacy to play AGAINST.

The magician should seek out enchantment objectives where the microstates of the system have a useful grain size but with only small energy differences between them.

For example a stationary dice shows just one of its six possible microstates and the transition energies between them remain fairly high, you cannot easily flip it by telekinesis. However whilst a dice tumbles vigorously the transition energies between its final microstates remain tiny.

The differences between microstates of the Boltzman-Gibbs and the Shannon equations illustrates this point, the information in a system may well exist at a grain size enormously greater than the molecular scale, so as a general rule of thumb throw information at a target rather than energy.

Addendum

Discworld magicians have two measures of magical 'strength' the Thaum and the Prime, which define units of magic in terms of information and energy respectively.

The Thaum defines the magic required to apport various objects. This implies a rearrangement of the information that previously defined their positions. Roundworld magicians call this retroactive enchantment.

The Prime defines the amount of magic required to shift a weight a certain distance, - (although Discworld magicians do not specify the timescale or the friction involved). We call this telekinesis, and regard the Prime as an energy measurement.

Now Discworlders use peculiar units, the appearance of a small white goose or three billiard balls for the Thaum, and the movement of a pound of lead one foot for the Prime. This results in the two scales having different numbers for equivalent phenomenon, much as we have 4.2 joules per calorie here due to an unfortunate choice of units.

However if the value of the Prime or the Thaum here on Roundworld equals 10^{-14} joules per second or its 10^7 bits per second information equivalent, we can calculate their Discworld equivalents.

For this we need to know the Discworld values of Planck's constant and Ubiquity.

To find these things we need to make just 3 assumptions:

1) $\dfrac{M_D}{L_D} = \dfrac{c_D^2}{G}$ Discworld also lies in a hyperspherical universe having the same gravity as Roundworld, but its own lightspeed.

2) $\dfrac{M_D}{L_D^3} = \dfrac{M_R}{L_R^3}$ The Discworld universe has the same density as the Roundworld one.

3) $\hbar_D c_D = \hbar_R c_R$ Discworld has a higher Planck's constant to balance its lower lightspeed of only 200mph.

Thus from 1 and 2 we can conclude that Discworld lies in a smaller universe, only about 3.25 thousand light years across (Roundworld light years), and weighing in at a mere 3.7×10^{33} kg.

From this and from 3 we can calculate U_D, the Discworld's Ubiquity constant.

Finally we can obtain $\hbar_D \sqrt[3]{U_D} = 9.17 \times 10^{-13}$
Thus the Discworld Prime equals about 10^{-12} joules/second and the Discworld Thaum equals about a billion bits per second.

Thus to put things in proportion:

$$\frac{\hbar_D}{\hbar_R} \frac{\sqrt[3]{U_D}}{\sqrt[3]{U_R}} \sim 50$$

So the Discworld's Thaum and Prime have about fifty times their Roundworld values.

That's enough to swat a mosquito, to cheat visibly at billiards, or to play havoc with a sensitive information system, like someone's brain.

No wonder Discworld also needs Narrativium to stabilise its fragile causality.

Chapter 5
The Spells of the Linking

$$d = \sqrt{x^2 + y^2 + z^2 + (ict_1)^2 + (i^2ct_2)^2 + (i^2ct_3)^2}$$

(3 Dimensional time, full form)

$$d = \sqrt{s^2 - (ct)^2 + (ct_i)^2}$$

(3 dimensional time, simplified form)

$$0 = \sqrt{s^2 - (ct)^2} \qquad\qquad 0 = \sqrt{(ct_i)^2 - (ct)^2}$$

(Entanglement)   (Superposition)

The whole idea of magic appears dubious to many modern scientific Roundworld minds because it implies connections between events that appear to have no possibility of causal connection. The first attempt to formulate some kind of mechanism or medium for the transmission of magical effects which didn't depend on religious concepts, began with the 19th Century magician Eliphas Levi. He hypothesised an 'Astral Light' as the medium of magic, and the provider of the magical link.

Writing of the magical link in *Magick*, Aliester Crowley noted that '**More failure comes from neglect of this than from all other causes put together.**'

Then in *Magick in Theory and Practise* he notes: -

'It is deplorable that nobody should have recorded in a systematic form the results of our investigations of the Astral Light. We have no account of its properties or of the laws which obtain in its sphere. Yet these are sufficiently remarkable. We may briefly notice that, **in the Astral Light, two or more objects can occupy the same space at the same time** without interfering with each other or losing their outlines.'

Most magicians have an intuition that some sort of 'Astral' plane or 'Etheric' link between events must play a part in magic, yet it remains as a sort of Elephant in the Room. It implies that magic depends on some massive but vague assumptions about reality, and that magic has few coherent theories about them.

A specification of exactly what 'The Astral' consists of and what it does, should lead to a systematic understanding of how magical links work and a reduction in the failure rate by working accordingly.

Adding together recent developments in the scientific ideas about waves and fields, elements of the fringe-scientific idea of Morphic Fields, plus some elements of magical theory, yields a realistic specification: -

The Astral consists of the Wave (and Field) Functions of Reality.

Now we normally perceive the Particle Function of Reality only. We perceive only the apparent doings of the particles which make up material objects; we cannot directly perceive the wavelike states that particles adopt between their doings, or the fields that mediate the forces between them, yet these largely define what they end up doing. However an examination of the peculiar phenomena that particles exhibit confirms that they must enter wavelike states, and indeed that they actually spend most of their time(s) there, and that they must also exchange field mediated effects.

The word 'astral' technically devolves from 'related to the stars'. However, since it has gained association with the occult we may as well hijack it completely and sharpen up its definition.

Magicians usually develop the intuition that material reality only constitutes a small part of the whole of reality and that reality has a much more extensive 'astral' part.

This intuition partly derives from the fact that we can imagine more than we can directly perceive. However the imagination does not in itself constitute the astral plane. The creative imagination arises at the interface of the astral wave function of reality with the sensitive particle structure in the brain.

The wave-particle aspects of the whole reality do not lead to the philosophical problems of the mind-body duality of the Cartesians. Nor do they lead to the metaphysical extravagances of the Platonic theory of archetypes. Reality exhibits particulate properties at the instant of interaction and observation, between such instants it exhibits wavelike properties. Our subjective experience of consciousness seems a bit odd and immaterial because it partakes of elements of both.

Now the wave-functions which describe what particles appear to have done between interactions do not make sense if we persist in describing reality as having three spatial dimensions but a single 'forward only' time dimension.

Two now commonly exploited physical phenomena become quite inexplicable using such a simplistic time frame, quantum entanglement and quantum superposition.

In entanglement, a certain instantaneous coherence often remains between events once in contact, no matter if they later get separated by arbitrarily vast distances. This seems best explained in terms of effects propagating backwards in time down the paths that led to their separation, to reset the starting conditions.

In superposition a particle can appear to have come out of a state corresponding to it having occupied several mutually contradictory states simultaneously in its past. In fact most of the basic chemistry of life, and inorganic chemistry as well, appears to depend on the ability of particles to have occupied mutually contradictory states, and then often to re-enter them immediately after observation or interaction. This seems best explained in terms of time having a sideways component as well to accommodate the superpositions. Thus the wave functions of the past and future of a particle consist of 'parallel universes' defined by a sideways separation in time rather than in space.

In the early days of excitement after the realization that electromagnetic radiation actually encompassed a huge spectrum of phenomena from gamma rays to x-rays, visible light and the radiation we can use in radar and radio, some occultists began to wonder if it could explain magical phenomena as well, and they tried to draw parallels between the astral and the 'luminiferous ether' which scientists (mistakenly) supposed would support the transmission of such electromagnetic radiation.

The luminiferous ether idea proved superfluous; radiation seems to travel quite happily through empty space on its own. Simple radiation on its own cannot account for magical phenomena because most of it doesn't travel round corners or through solid objects very well, and it doesn't persist, it flies from emitter to receiver and then disappears.

Yet despite the abandonment of the luminiferous ether concept, theorists have retained the idea of various 'fields' which permeate all space.

We now know that the dual wave-particle nature of light also applies to all the particles of matter as well. They too have a wave-particle nature but generally with much smaller wave functions. The phenomena we call particles, like the electrons and protons of ordinary matter, actually seem to consist of the events where the wave functions from their pasts and futures overlap at the instant of their interaction.

Thus on close analysis the universe appears disturbingly weird and illusory, the particle reality, full of the material objects that we usually perceive, recreates itself every instant out of a much larger wave reality that we cannot directly observe. Only the comparatively small wave functions of matter particles keep the universe re-manifesting in more or less the same form from one instant to the next. Energy particles that have larger wave functions tend to fly around even less predictably.

Now a mere century and a half since Eliphas Levi proposed a mechanistic rather than a theological basis for magic, we approach Crowley's dream of a systematic account of the properties and the laws which pertain to the astral.

This involves some deceptively complicated mathematics; but DON'T PANIC, it yields easily understandable results.

Ouranos

Chaometry: Part 2

The so-called 'imaginary' numbers, the square roots of various negative ordinary numbers, provide useful tools for describing 'unobservable' quantities. 'Unobservable' in this sense means that whilst we cannot get hold of them in any sort of concrete sense we acknowledge their effects and allow for them. Then at the end of the calculation we get rid of them to find out what happens in the observable world.

Thus these imaginary numbers make ideal tools for describing the effects of the astral or wave reality on the particle reality or 'material world'.

As Special Relativity predicted and as experiment has repeatedly confirmed, time slows down on an object as its speed increases. At lightspeed its onboard time stops completely. So from the perspective of the object, the journey becomes instantaneous, time becomes unobservable, and the spacetime distance, d, experienced by the object drops to zero.

The use of the imaginary number i, the square root of minus one, models this accurately:

$$d = \sqrt{s^2 + (ict)^2} \quad \text{Minkowski}[20]$$

(This shows the simple Pythagorean formula,

$$a^2 = b^2 + c^2$$

or, $a = \sqrt{b^2 + c^2}$ adapted to include time.)

However to include time, t, we need to first multiply it by lightspeed, c, to convert it to the same units as space, s, and then to multiply it by i, to account for its 'unobservability'.
The procedure of squaring (ict) will remove the imaginary number i, converting it back to minus one, so we end up with:

$$d = \sqrt{s^2 - (ct)^2}$$

So for a journey at lightspeed the experienced spacetime distance, d, drops to zero. Now only the waves, which appear as photon particles when you stop them, can manage to travel that fast. All mass carrying particles have to travel slower. However the various force carrying fields, nuclear, electromagnetic, electroweak, and gravitational, do also appear to propagate at lightspeed.

The first of the Spells of the Linking shows all three dimensions of time:

$$d = \sqrt{x^2 + y^2 + z^2 + (ict_1)^2 + (i^2ct_2)^2 + (i^2ct_3)^2}$$

This Spell / Equation shows all three spatial dimensions, denoted x, y, and z plus the three dimensions of time. For the sake of simplicity it does not show the two curvature 'dimensions' of space and time which make no significant difference in weak gravity conditions.

'Ordinary' time as defined by the direction of measurement of elapsed time or the time associated with travel velocity appears as t_1 and it undergoes the conventional multiplication by c and i. The two other dimensions of time, denoted t_2 and t_3 appear multiplied by c, and also by i^2. Now this may seem a curious step to take, but if an observer measures elapsed time, arbitrarily designated t_1, then from that observer's perspective, t_2 and t_3 remain doubly unobservable, hence the use of i^2. This has the effect of describing the two 'sideways' dimensions of time as a sort of plane of pseudo-space (unobservable space) from the perspective on the observer.

If we then roll up the three spatial dimensions into a simple spatial distance, s, and roll the two sideways dimensions of 'imaginary' time into a single component, t_i, and square out all the i factors, then the equation can appear in the more useful abbreviated form:

$$d = \sqrt{s^2 - (ct)^2 + (ct_i)^2}$$

In other words, the spacetime separation between events becomes a function of the spatial distance minus the time, plus the imaginary time. This yields many solutions where the spacetime distance equals zero and the events lie entangled together, or in a condition of magical linkage.

Two of the zero case solutions provide illustrative examples:

$$0 = \sqrt{s^2 - (ct)^2} \qquad \text{(Entanglement).}$$

Entanglement occurs where a 'light path' exists between events separated by space. Note that because time enters the equation in squared form, both future (time positive) and past (time negative) give the same result.

$$0 = \sqrt{(ct_i)^2 - (ct)^2} \qquad \text{(Superposition).}$$

Superposition occurs where probabilities in the same place lie separated by 'sideways' imaginary time.

Entanglement and superposition represent two extremes of the same phenomenon, as a continuous range of zero solutions exist with various admixtures of spatial distance, s, temporal separation, t, and imaginary time separation, t_i. It seems that none of the individual separations ever has an absolutely zero value in itself, probably because of an underlying quantisation of spacetime. Thus entanglement always has a superposition component across a small imaginary time interval and this makes it a probability based connection rather than a causal one. Similarly the imaginary time separation of superposed states at apparently the same spatial location does have an unobservable but inferable pseudo-spatial displacement.

The imaginary time plane (which acts like a sort of pseudo-space) lies orthogonal to ordinary time (and also in some un-visualise-able sense, orthogonal to all three dimensions of space as well). It thus acts as the hidden variable which makes the universe appear to run on probability rather than determinism. Probability lies at right angles to time as they say.

Now philosophers and metaphysicians may argue that if imaginary time supplies the hidden variable which actually controls what the quanta do, then we inhabit a causal rather than a probability based universe but we just cannot see the causal mechanisms at work because they remain unobservable *in principle* as well as in practise. Hence they may argue, as Einstein did, in favour of a strictly causal universe, and that quantum physics remains incomplete.

However the hidden variable function of imaginary time plane probably acts randomly as, somewhat paradoxically, only randomness can create the individually unpredictable but statistically regular effects that we observe. Moreover, in a three-dimensional time frame, superposed events can lie in any direction on the imaginary plane from a point designated on the measurement axis t_1. We cannot in principle tell in which direction, and that may play a part in affecting what happens, and events may select their directions randomly.

Thus despite that Einstein insisted that *god (the Universe) does not play dice*, it appears that she does, and moreover she throws them where nobody can see them – in the imaginary time plane, which thus supplies a chaotic hidden variable.

So can we make any use of random events and the probabilistic connections between them?
Well random quantum events do at least lead to statistically reliable outcomes in bulk, leading to the sort of apparent causality on which classical science depends. However the magician usually has more interest in the anomalies and the exceptions than the rules.

We can visualise the imaginary time plane spreading out sideways like a space pierced in the middle by the thread of the 'real' time axis, even though the real time axis only arises by stringing together remembered instants of the past or expected instants of the future. The real time axis we thus construct may appear 'straight' but it has no more straightness than the path taken by someone stumbling in total darkness, for it meanders sideways in directions previously designated as imaginary, making them 'real' as it does so.
Yet the further from the central thread an area lies, the less likely that the thread will pass through it and make it real. Thus the magician needs to interact with imaginary time sparingly, its closer reaches encompass what can probably happen, its further reaches encompass the far larger reaches of what probably won't.

Now information processing machines, of which the human brain provides the most powerful currently available example, can create their own waves in imaginary time.
A visualised or imagined event can have a similar effect on the imaginary time plane as the probability wave function of

a material event, because it too constitutes a wave-particle event.

Physical contact or 'line of sight' provides the best opportunities for a magical link.
Here the actual visual sight of the target does not matter except insofar as it may help with the visualisation, and 'line of sight' in this context includes any light-like path of any spatial distance, irrespective of obstructions. Probability waves act rather like neutrinos and pass through solid matter largely unobstructed. Thus 'line of sight' in this context pretty much reduces to real time simultaneity for all practical purposes, the best magical links occur between simultaneous events as reflected in the second of the magical link equations:

$$0 = \sqrt{s^2 - (ct)^2}$$

Note that imaginary time separation does not appear here as it has a very small value. The magician can thus establish a link with the higher probability wave states that lie close to the target in imaginary time, and which represent its immediate pasts and futures. In the absence of physical contact or a visual line of sight to the target, the magician needs to visualise the target in its actual condition at the moment of the desired magical link to achieve the highest quality of link. Remembered images provide a poor substitute, and don't forget that most video and tele-visual feeds have undergone a time delay. A live telephone conversation with some covert magical intent may often give a better link.

Working magic across real time intervals presents some special problems because it necessarily involves a more or less equal amount of imaginary time as well, as the third of the link equations shows:

$$0 = \sqrt{(ct_i)^2 - (ct)^2}$$

Here the spatial interval effectively reduces to negligible again for work conducted at less than astronomical distances, only imaginary time can cancel the real time separation. Now the larger the imaginary time separation becomes, the more the probabilities in it multiply.

Fields.
In the conventional quantum model the fundamental nuclear, electromagnetic, electroweak, and gravitational fields supposedly arise from the action of so called 'virtual particles'. These mediate the rather ill-understood forces in the nuclei of atoms in the case of the strong force, and phenomena like electric and magnetic fields according to standard theory. Particle physicists like to consider that gravity works in a similar way although relativity physicists prefer to consider gravity as a curvature in spacetime. Now these virtual particles differ from real particles like photons in that you cannot in principle catch them or observe them directly and they have rather odd properties like an apparent indifference to solid objects, through which they pass unobstructed. Although we can only measure the forces between objects affected by fields, the presumption exists that such fields have an all pervasive quality. Real particles such as photons then appear in the theory as wavelike disturbances in the underlying ubiquitous fields. Thus for example the theory implies that gravitational fields arise from the actions of virtual gravitons. These remain unobservable except as the 'force' of gravity. However the sudden acceleration of big enough lumps of matter, as perhaps provided by a couple of neutron stars colliding, should produce real graviton particles which we could in principle detect as a passing gravity wave. Yet we could probably never detect real gravitons individually, because of

their weakness. So in a sense, field theory recreates the older luminiferous ether idea and it invokes other 'ethers' for the other forces.

Now the Hyperspin Eight Dimensional (HD8)* hypothesis has no need of virtual particles. Instead it extends the general relativistic idea of a gravitational field as a spacetime curvature to model the other forces in the same way, and it uses extra (temporal) dimensions in spacetime to support the other curvatures.
(*See The Apophenion)

In either model the fields of the fundamental forces have a virtual quality, either from the virtual particles or from the imaginary time components of the higher dimensional curvatures. Both gravitational and electromagnetic fields have universal pervasiveness through empty space or solid objects, and both have unlimited range attenuated only by the inverse square of distance. Gravity fields appear to act attractively only, whilst the other fields arise from positive and negative charges which can attract or repel. This apparent asymmetry arises because nuclear, electromagnetic and electroweak charges have imaginary time components, so from an instant of real time observation these charges can point in 'forward' or 'backward' in directions orthogonal to the real time axis. Gravity does actually work repulsively as well, as you can observe if you reverse the real time axis. Try watching a film in reverse, things fall up.

All objects have their own measurable gravity fields but objects with no overall charge do not exhibit measurable electromagnetic fields because the internal positive and negative charges both create fields which have an equal and opposite effect on any measuring apparatus. The same probably applies to the strong nuclear charges and explains our inability to detect these awesome forces outside of the nuclei of atoms.

Just how much information such fields can support or transmit remains an open question. However if all the components of an object project various fields, then in principle these fields carry a complete specification of the object. They carry its entire morphology (in the form of its wave functions) in some sort of virtual form. Now such fields attenuate with the inverse square of distance, but this makes little difference in terms of information, within wide limits a radio receiver gets the same message from a transmitter no matter what the distance.

Such morphological fields or 'Morphic Fields' to borrow a term coined by Rupert Sheldrake [21] , cannot transmit signals in the normal sense of a modulated stream of information, they can only convey the 'configuration' of the transmitter in terms of its wave functions which remain probabilistic. Only receivers with a very similar configuration can pick up such morphic fields, and even then the resonance between transmitter and receiver remains probabilistic rather than directly causal.

Nevertheless this apparently weak effect probably accounts for all the so-called physical laws and physical constants and various other arbitrary conventions which more or less seem to govern and structure this universe.

Yet such a system remains partly chaotic and it contains enough slack to allow for adventitious developments such as living organisms and magic.

Thus the virtual fields and wave functions associated with what we think of as 'matter' actually keep matter behaving in a more or less regular manner. All physical laws imply an underlying conservation principle as Emmy Noether* [22] realized, and the virtual fields and wave functions act as the agents of that conservation. However when novelty arises through some loophole in the system they will tend to conserve that also.

*(Einstein Hilbert and others viewed her as the most important woman in the history of mathematics)

Now fields almost certainly consist of the wave functions associated with particles, as Schröedinger originally surmised.

Thus The Astral = The Ether = The Wavefunctions = The Fields, and this part of reality performs the function of so-called Morphic Fields, but with one exception to the Sheldrake theory, Morphic fields do not survive the disintegration of the particle forms associated with them. The 'Persistence of the Past' does not actually occur, reality forgets; it has to, or it would get clogged up with all the minutiae of what it has done and become inhibited from doing anything else.

So to recap briefly: -

1) Reality exhibits a Particle - Wave duality. To a human observer the particle aspect of reality appears real and the wave reality appears virtual.

2) Wave functions spread out in space and time to create fields.
Fields do not carry information as such, but rather 'form' or morphology.
Such fields maintain physical laws and constants, they act as conservation principles in the sense that Noether implied.

3) Fields propagate at lightspeed. (But forwards or backwards in all three dimensions of time). Thus we can at least work out some of the rules of their operation.

4) Fields correlate by affinity. Fields 'from' an event only affect other events that have a very similar form or shared characteristics.

5) No Presence of the Past. Spells and telepathy cease once the person projecting them stops doing so. Gods who run out of followers cease to have any effect. The Morphic field of dinosaurs has ceased to exist. Protons have maintained a constant electric charge since time immemorial because we have always had plenty of them about to keep it that way.

Whilst nobody disagrees that the mathematical formalisms of quantum mechanics have very high predictive powers, at least statistically, very little agreement exists as to what such formalisms actually mean for our view of reality. Thus a vast metaphysical argument rages about the possible interpretations. Just what do they mean in terms of whether we should consider the universe to run on cause or chance, or with consciousness or without, or on principles that we can understand or on ones that we cannot understand, even *in principle?*

The preceding exposition has used the Maximalist Hypothesis.

In the absence of any definitive evidence to decide between YES or NO for any of the following six questions we have opted to say yes to all possibilities. As such it provides yet another 'unique' interpretation of quantum mechanics, of which on this basis alone, 64 possible combinations may exist.

Indeter minacy?	Wave function Real?	Multiple Histories?	Hidden Variables?	Collapsing Wave functions?	Obser ver Role?
YES	YES*[1]	YES	YES*[2]	YES	YES

(*[1], Wavefunctions remain unobservable but have real effects.)

(*[2] Imaginary time acts as a chaotic hidden variable.)

Each of the 64 possible interpretations might lead to a different metaphysic, and each to a slightly different theory of magical link. Chaoists may care to experiment with the consequences, and indeed perhaps only such magical experimentation rather than conventional scientific investigation may allow us to rule some of them out.

However the quantum model in general, regardless of interpretation, does allow us to make some educated guesses backed up by empirical experience.

In terms of general principles, for an effective link, the magician needs: -

A) Affinity. The magician needs a 'mental' model that corresponds as closely to the target as possible. This model does not have to exist in conscious form, indeed as Chapter 6 shows, the magician should avoid giving the conscious representation of the target too much attention, but a high quality subconscious representation needs to exist for an effective link.

B) Simultaneity. As ('astral') wavefunctions and ('etheric') fields propagate at lightspeed the magician can potentially have a more or less instantaneous link to anywhere on a planet such as this one, which has a diameter of only about a twentieth of a light-second. Thus for planetary scale work we can effectively discount spatial distances. However if the magician attempts to form a link to events with a significant real time difference then only imaginary time can provide a conduit for a link,

$$0 = \sqrt{(ct_i)^2 - (ct)^2}$$

However, the more use the magician makes of imaginary time the greater the uncertainty becomes, primarily because we have no reliable faculty for perceiving directions or amplitudes in the imaginary time plane. Thus divination gets progressively more precarious for events that lie separated in time from the magician.

Hence the old dictum, 'Divine Short!'

On the other hand we do have some facility for structuring the imaginary future and for re-structuring the imaginary past, namely the faculty of willed imagination. Now the wave functions tend to spread out and multiply across time but this works to the magician's advantage in Enchantment because it allows for the possibility of upping the chances of lower probability events by reinforcing them with some artificially generated affinity, for example by visualising them happening.

Hence the other old dictum, 'Enchant Long!'

Most magicians would probably prefer it the other way around, and to have the ability to divine events way into the future and to enchant for fast effect, but wishful thinking has its limits which we have to recognise.

A consideration of the above principles allows for a realistic assessment of the chances of success with various types of magical link.

Magical Ritual Part 5:
The choice and use of Magical Links.

A consideration of the typology of magical links leads to their classification into seven categories, some of which overlap. We present them in order of effectiveness.

1) Physical Contact and Line of Sight.

2) Links to Future Selves.

3) Visualised Links.

4) Telesmatic Links.

5) Real time links across occluded space.

6) Real space links across occluded time.

7) Occluded links across space and time.

Physical contact or line of sight to the target provides the best possible form of magical link as it minimises the imaginary temporal separation and it also provides the opportunity to visualise the target in real time. Some magicians refuse to work with anything else.

Links to future selves remain a rather undervalued resource, particularly in divination. Of all the things in the future that could act as a source of an answer to a question, the future state of the magician's own self when it has the answer offers the best possibilities.

Dowsing for example seems easiest explained in terms of the dowser basically asking 'what future experience will I have if I dig here', rather than supposing that the effect derives from mysterious emanations from buried resources. Virtually all accurate premonitions seem to relate to events that have a significant emotional impact on the diviner. Some magicians make a point of trying to visualise themselves at some point in the future by when they will probably have an answer to their divinatory questions. Some will only divine for another if that other person solemnly agrees to notify them of what actually eventually happens.

Visualised magical links often have to suffice in the absence of opportunities for physical contact or line of sight links. Remembered images from actual contact work much better than recorded images or even audio-visual transmission records which contain only a fraction of the information gleaned from actual contact.

Telesmatic links come in at a very poor fourth. Personal possessions rarely remain in any sort of state of entanglement or superposition with their owners for long. However prized possessions of a singular or unique nature may still occupy a place in the minds of their owners and this provides a link of sorts, if a tenuous one. Skin-flakes, hair and fingernail clippings do however carry a unique genetic fingerprint and hence a morphic link to their origins. Nevertheless such telesmata only have a value as magical links to the living. The dead no longer have the capacity to send or receive anything.

Necromancers actually create 'spirits' within their own heads and delude themselves otherwise.

Real time links across occluded space generally require the use of visualisation or telesmata although an additional real-time connection via an electronic communications network often helps.

Real space links across occluded time suffer from the usual corrosive effects of proliferating probabilities in imaginary time if the magician attempts divination. In enchantment for an effect to manifest in the future the magician usually aims spells at the target in the present. However in retroactive enchantment where the magician attempts to modify the past, links become more tenuous with the passage of time.

Establishing links across both occluded space and time presents almost insurmountable difficulties and few serious magicians attempt to divine or enchant events removed from them in both dimensions.

Chapter 6
The Spells of Impractical Magic

$$P_\Psi = P + (1-P)\,\Psi^{1/P}$$
(Spell)

$$P_\Psi = P - P\,\Psi^{1/(1-P)}$$
(Antispell)

In the Chaoist view, magic works by influencing probability. Any event has a probability of occurrence somewhere between 1 (for certainty) and 0 (for impossibility).

Thus a fair coin toss has a probability of 0.5 for coming up heads and a fairly thrown dice has a probability of about 0.16 of showing a six.

However magicians tend to regard such odds as 'potentially improvable' and they tend to regard nothing as either completely certain, even though it may appear about 0.99999 so, or as completely impossible either, although 0.00001 probabilities do seem challenging to work with.

Daunting probabilities often appear for complex tasks like winning in war or business or love from a weak initial position; however all of the steps involved in winning may not actually look so impossible individually. The probability of throwing a dozen heads consecutively comes out at a desperate 0.00024, so only one person in four thousand might achieve it on average. However each individual step has a 0.5 probability and a magician with any sense will choose to attack such steps individually rather than go head to head against such formidable odds.

In practise the difficulty of manipulating probability by magic varies with the probability itself. A given amount of magic makes a proportionally more useful difference to an event of medium probability than to a very high or a very low probability event.

This relationship finds expression in a Chaometry tool called the second equation of magic: -

$$P_\Psi = P + (1\text{-}P)\,\Psi^{1/P}$$

This relationship derives from simple principles to give a very accurate estimate of the expected probability distortion for any act of magic. It has a particular value in illustrating the sort of enchantments and divinations that we should regard as impractical and not worth the effort and those which may stand a chance of success.

It states that the probability of achieving something by magic, P_Ψ, equals the probability of its 'natural' occurrence P, plus the rest of the probability, $(1 - P)$, multiplied by the amount of magic used Ψ.

Magic acts on the remaining probability, $(1 - P)$.

However we need to modify magic factor Ψ by the power of the reciprocal of the natural probability, $^{1/P}$, to reflect the increasing difficulty of affecting events of low natural probability. As the natural probability drops $^{1/P}$ becomes larger, so multiplying Ψ by itself this number of times tends to reduce it, as Ψ never exceeds 1.

Chapter 7 shows the first equation of magic which gives the formula for calculating Ψ, the amount of magic, on a scale of nought to one.

In graphic form this equation leads to the Tripod of Stokastikos image shown below.

We cannot show the equation as a two-dimensional graph because it has three variables of P_ψ, P, and Ψ, so it appears as a 3D structure with the surface of the net rather than a simple line representing its solutions. Additionally the P axis conventionally lies projected out of the page from 1 to 0, to make the net easier to visualise.

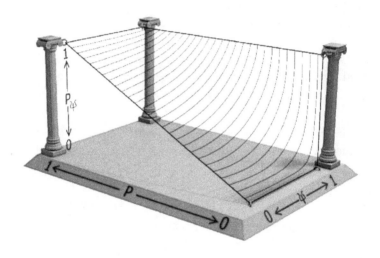

Figure 7. Tripod of Stokastikos

Now the Tripod has some peculiar features. The height of the net above the base represents P_ψ, the probability of achieving something by magic.

The sloping left hand side of the net represents the rather blindingly obvious, where the amount of magic used equals 0, (either because none gets used or hopeless technique ensures total ineffectiveness), then the probability remains unchanged from the natural probability. The right hand side of the net shows that no amount of magic makes any improvement to a probability of 1. All the interesting results lie between these extremes.

Note the lines that curve up on the net from the left-hand side. These represent the effects of magic on various probabilities; all of them eventually end up on the top right hand edge of the net with a P_ψ value of 1, so a sufficient amount of magic can in principle raise any probability to certainty. However there seems little point in expending a lot of magical effort to reinforce desired events of already high natural probability. The bottom line on the net, the front edge of it, shows the effect of magic on an event of natural probability zero, which corresponds to an impossible event. It remains impossible until the magician applies $\Psi = 1$ grade magic to it, whereupon it suddenly becomes a certainty. In practise such extreme acts of magic prove extraordinarily difficult, particularly for large-scale events where the belief and magical link may become a problem. (See Chapters 5 and 7).

For practical purposes the central area of the net on the tripod shows the most interesting opportunities for magical attempts to distort probability.

Here for example we see that if an event has a natural probability of 0.5 then the application of Ψ 0.5 pushes the P_ψ value up to just 0.625, but the application of Ψ 0.8 pushes it up to a respectable 0.82. Note that these figures apply equally well to divination where P then means the probability of arriving at the answer by chance and P_ψ means the probability of arriving at it by using magical divination.

In a probability-based universe, enchantment seems preferable to divination. As neither usually manages to achieve certainty, a failed enchantment merely provides an opportunity for a revised attempt, but a wrong divination can lead to disaster if acted upon.

The second of the spells of impractical magic represents an inversion of the first and it shows the probability of preventing something by magic by decreasing its probability of occurrence. It too has the curious right-angled feature in that it shows an extreme act of magic, $\Psi = 1$ turning a certainty of occurrence, $P = 1$, into a zero probability.

Magicians generally try and avoid working with certainties or impossibilities because they cannot modify them, they can only reverse them completely in exceptional circumstances.

Psi Ψ, Means the amount of magic that the magician can bring to bear on a situation. Where a conjuration achieves $\Psi = 1$ it brings a whole Prime Π or Thaum Θ into play. (See chapter 4).

Thus a numerical equivalence exists; $\Psi \equiv \Pi \equiv \Theta$.

It seems that in practise a group conjuration for some objective does not have a cumulative effect except in that it may allow or provoke some participants to put in an exceptional effort. The effect of a group does not exceed the effect of the most effective participant. However just who achieved that usually remains an unasked question for reasons of professional etiquette amongst peers, as with firing squads.

Magical Ritual Part 6: chaometry (iv)

Choosing achievable objectives for the conjuration phase of a magical ritual involves a careful consideration of the art of the possible. In any conjuration the magician seeks to encourage something happen, or to find out something previously unknown or uncertain. Enchantment and Divination represent the direct route to the exercise of magical will or perception but the magician may also use indirect methods such as Invocation, Evocation, or Illumination, particularly in the pursuit of more complicated objectives.

In every case the magician needs to consider the Chaometry: -

(i) Subtlety. Do the required changes lie within the energy, entropy, and informational indeterminacies of the system?

(ii) Linkage. Does a reliable magical link exist?

(iii) Practicality. Does the conjuration seem likely to achieve a useful distortion of probability?

Considerations of practicality depend very much on the amount of magic, Ψ , required.
Chapter 7 discusses the requirements for making it available. However Chaometry considerations usually rule out most of the things that many people fantasise about achieving by magic as simply impractical by direct Enchantment or Divination.

Magicians therefore often elect to work towards an objective by the indirect methods of Invocation, Evocation, and Illumination. A brief discussion of each of these in terms of Chaometry now follows: -

Chapter 3 addresses Invocation in some depth. Basically Invocation of some Chaomera or god-form allows the magician to enter gnosis more readily by providing a path of emotional egress into a previously established 'exalted state' of hyper-focussed excitation or quiescence as Chapter 7 explains. Secondly it provides access to previously established belief structures which help the magician overcome disbelief in the possibility of the desired outcome. Thirdly the very act of Invocation helps to delete a lot of contrary conscious mind chatter and allow the intention to reach the more powerful subconscious. Thus in Invocation the magician uses 'Sleight of Mind' tricks to overcome the normal consensus straitjacket mindset of our culture that denies the possibility of magic.

Evocation makes use of similar tricks except that the magician conceptualises an evoked entity as a sort of semi-separate intelligence or servitor budded off and departmentalised within the subconscious and delegated to perform various tasks autonomously, rather than as something to enter wholeheartedly into identification with, or possession by. Magicians frequently use Invocation to achieve communion with a god-form and then Evocation of an entity of some sort to automatically work on a task of a germane nature.

Thus a god may command a legion of minor spirits and demons: to use an old fashioned terminology

Such 'Fire and Forget' magical missiles work all the better when the magician forgets about them.

So in the Chaoist view, three somewhat arbitrary and overlapping delineations exist, simple sigils for well defined and simple objectives, evoked entities for tasks requiring a certain amount of antonymous intelligent action, and gods for the complicated objectives like life changing inspirations or additions or deletions to the personal pantheon of the selves.

However as the old axiom goes, never Invoke what you should have merely Evoked, the magician should never allow minor desires to become obsessive and interfere with the pursuit of High Magic. See chapter 8, and what follows; -

Illumination techniques frequently involve the whole spectrum of conjuration from simple spells to Evocation and to Invocation. The magician needs to consider what counts as real advancement, what balance of achievement, power, wisdom or knowledge seems best on a personal or a species or a cosmic level? Try for example formally Invoking Death. What sort of funeral do you want? What do you want your species to remember you for? What do your selves want to remember each other for?

There seems little point in practising magic unless you want to do at least a few extraordinary things with this incarnation.

Chapter 7
The Spell of Practical Magic

$\Psi = GLSB$

Traditionally the Witch or Magician received upon initiation the following exhortation: -

To Know, To Will, To Dare, and To Keep Silent.

These so-called Powers of the Sphinx or the Witches Pyramid have remained the subject of much speculation and varied interpretation over the centuries.

Obviously 'Knowledge' remains important at any time. The pursuit of magic also calls for a certain amount of 'Willing' and 'Daring' particularly in cultures that forbid it. Plus of course 'Keeping Silent' often kept you and your friends and colleagues from the inquisition's pyres in days past.

However these seem rather trivial interpretations, and in terms of contemporary magical theory and practise we can now give a deeper and more technical insight into these esoteric ideas in terms of how they relate to the actual sleight of mind manoeuvres necessary for magic.

Of all the things a magician needs To Know, the target of the intended magic remains supremely important. Without a serviceable magical link the magician can accomplish nothing. Much of the lore of historical magic confuses the idea of the magical link with the idea of the symbolic attributes of the target in terms of its supposed astrological, elemental, and alchemical properties etc. These do not act as effective substitutes for a magical link, but rather they relate to the idea of Silence in which the magician uses some symbolic representation during the actual working to distract the conscious mind whilst the subconscious mind, which has to contain a serviceable magical link, effects the magic. Thus before any magical action the magician needs to acquire an accurate and as up to date mental model of the target phenomenon as possible. The basic techniques of magic remain fairly simple but they have no effect if you shoot blind or at a largely imaginary target.

Magical 'Will' remains a much misunderstood topic. Conscious desire rarely has much parapsychological effect no matter how fervently the magician tries to amplify it. In fact conscious 'Lust of Result' often inhibits the use of magic. If conscious desire had much real effect then we would inhabit a universe in which wishes came true to an alarming and frighteningly immediate effect.

Conscious desires tend to have a habit of only coming true once the conscious mind has forgotten about them, and this observation provides the key to the magical use of 'will'. Many people get what they want eventually but often too late, the magician cultivates sleight of mind effects to speed things up a bit.

To perform magic effectively the magician uses will to enter a highly focussed state of consciousness which doesn't actually include conscious desire at all, but which focuses on an abstract representation of desire instead. This then provokes the much more powerful subconscious mind to activate the actual desire and the magical link that it has stored. Thus when conjuring, the magician seeks a highly focussed 'Gnostic' form of concentration in which an abstract representation of desire, rather than a direct representation, acts as the object of either deep meditative concentration or ecstatic emotional arousal. Thus the magician needs to use Will to access the mental condition of Gnosis [23] ; simply using it to hypertrophy conscious desire or lust of result has very little magical effect.

Daring implies some sort of belief in the possibility of anything you attempt to achieve, or at the very least some suspension of disbelief in its impossibility. Nobody ever attempts anything that they consider truly impossible, except in the hope of embarrassing someone else who asserts otherwise.

Most of the paraphernalia of magic, all the wands and the robes and the ideas in the books, serve mainly to reinforce the magician's intuition that magic might just prove possible and therefore looks like a potentially worthwhile investment of belief, in cultural climates which usually remain hostile to the idea.

Some religions reserve magical powers for their deities only, but a few grudgingly allow for 'miracles of faith'. The classical (non-quantum) sciences generally regard the pursuit of magic as complete madness and delusion, a hubristic exercise that cannot possibly yield any result because the laws of causality forbid it.

Thus those magicians who have abandoned theological explanations for magic have mostly opted for either a studiedly irrational Romance of Sorcery or a Quantum basis for their belief system, or sometimes a quirky mixture of both.

Keeping Silent also has an inner esoteric meaning. It means stopping the conscious inner dialogue by some sleight of mind to allow the more powerful unconscious or subconscious to go to work.

All the spells in books work by analogy. They look like some symbolic representation of what the magician wants rather than the thing itself.

Most of the books however fail to make clear that the magician needs to focus as strongly and 'gnostically' as possible on the symbolic or analogical representation and to completely forget about what it represents consciously.

Conscious desire frequently proves fatal to magical intent because it always surrounds itself with doubt in the inner dialogue. Only desires which remain subliminal but intense usually bear fruit.

After conjuring the magician needs to avoid thinking consciously about the desire by a forceful turning of awareness to other matters. So-called banishing rituals can prove useful here, as indeed can any form of distraction including deliberately bursting out laughing and then thinking of something else.

All those tables of correspondences about the supposed occult qualities of various phenomena, mineral, vegetable, animal and metaphysical, do not really constitute very high grade magical knowledge. They have some uses in 'ensigilising' desire by offering symbolic and analogical substitutes. They may also have some value in deciphering divinations where the answer presents itself in symbolic form rather than by direct prescience. However in each case only meanings and correspondences that the magician has already accepted will prove of any value, so many contemporary magicians make up their own idiosyncratic ones.

Chaometry Part 4.

The four mental 'manoeuvres' of establishing 'Gnosis', 'Link', 'Subliminalisation', and 'Belief' create the Magic, Ψ, available in any conjuration and a shortfall in any one reduces the entire effect dramatically.

You can only build a pyramid as high as its shortest face as they say.

These considerations find expression in The Seventh Spell, The Equation of Practical Magic;

$$\Psi = GLSB$$

In this equation we get the total amount of magic available (Ψ) for any conjuration by taking the values of achievable Gnosis, Link, Subliminalisation, and Belief, all on a scale of 0 to 1, and multiplying them together.

The results of this look harsh, but they realistically model the difficulties involved.

Of course a zero value for any one of the factors results in a zero result, but even if the magician achieves 0.5 on all of them the overall result comes out at a hopeless 0.0625, or a mere sixteenth of the optimum of $\Psi = 1$. Half measures will plainly accomplish not half a result, but probably none at all.

However if all factors reach about 0.85 the magician can achieve $\Psi = 0.5$ which as the sixth equation shows, can make a useful difference to probabilities of medium values.

Thus magicians really need to give their conjurations everything they've got, and this often proves very hard and exhausting work.

Because of this, successful magicians tend to fall into two classes, 'Crisis Magicians' who only seem to get results when attempting something out of desperation, and 'Persistent Magicians', who spend a lot of their spare time practising at their gnosis, visualisation, sleight of mind, and belief techniques.

Towards both these types, but mainly the latter, we address the following section on general practical techniques.

Magical Ritual Part 7.

In working towards some kind of conjuration the magician needs first to have built up reserves of belief, then to have established a magical link particular to the desired objective, then to have somehow subliminalised or ensigilised the desire, and then finally the magician needs to launch the conjuration with gnosis. Thus we address these aspects of building the 'Pyramid of Ψ ' in that order.

In the Chaoist paradigm beliefs remain tools rather than ends in themselves. We should own our beliefs, they should not own us. Someone with sets of mutually inconsistent beliefs kept in separate compartments actually has a more useful toolbox than someone who insists on a single consistent set of beliefs.

Mutually inconsistent belief systems also provide a useful defence against paranoia and a mechanism for selectively ignoring the unwanted beliefs that come with most systems, one may as well have the best of both worlds as it were.

Nothing has absolute truth, anything may prove possible.

In building up belief in the possibility of having a 'magic-capable self' the apprentice magician will find it useful to acquire various physical tools and markers whose conscious manipulation and use on the physical plane forms a link to developing subconscious powers and beliefs. We do not so much do what we believe, as believe what we do. Ritual implants belief. Religions exploit this effect with breathtaking cynicism; we can apply it to ourselves for our own purposes.

A cupboard full of wands, robes, rings, amulets, crystals, elemental groundsleves [24] and other technical apparatus does wonders for magical belief, but only if you invest belief in them first by going to some trouble to obtain items of real personal significance and then practise with them regularly. Handmade or at least self-designed items work far better than purchased items unless these have highly unusual or highly significant origins.

As magic remains a minority belief system in most cultures magicians should conceal their ideas from hostile civilians and outwardly remain humorously dismissive of the whole concept, one may as well forestall unproductive criticism by supplying it oneself, or pass off one's interest as merely scholarly.

Magical Links have received extensive treatment in Chapter 5, but let it suffice to add that the magician needs to assume as little as possible, to hunt for information like a hawk, and to learn to visualise to the point of hallucination. The magician needs to have built in a serviceable magical link within the subconscious before conjuration begins, but to give it no conscious attention during the conjuration.

Factor 'S'. No single English word quite captures the process by which the magician creates a Symbolic representation, or some kind of Subliminalisation or Sigilisation or Spell or trigger for Subconscious action of the intent of the conjuration.

The libraries of magic overflow with formal recipes for creating such ensigilisations but in practise anything will serve so long as the magician goes through the process of making it or understanding its creation.

To make a physical sigil, simply make a more or less abstract representation of desire with some kind of clay or some unusual arrangement of mundane objects.

To make a verbal spell, simply mangle up the vernacular language expression of desire in some inventive way, or transliterate it into some language that remains incomprehensible to you.

To make a visual spell simply make a stylised graphic of the desired result and keep modifying it till it transforms into a completely abstract design.

To make a spell that you can visualise take either a graphic representation or a written statement of desire and mutate and reduce it until you have something abstract and simple enough to visualise

The Sigilised, Subliminalised, Spell form of the desire turns into the actual focus of the Gnosis during the conjuration.

Gnosis depends on achieving either a condition of extreme mental focus by withdrawing the attention from all but one thing or on achieving a condition of extreme mental focus by hyperactively creating awareness of one thing. In practise the same condition results whether you approach it by quiescent methods or ecstatic ones.

Many methods and techniques exist. Yogic type approaches to mental quiescence and focus characterise one way, Tantrik type methods involving ecstatic degrees of arousal by eroticism or anger or some other strong emotion characterise the other.

Magicians should master at least one technique of each type. The Tantrik style of gnosis seems better reserved for in-temple work or at least in-private work. However a competent magician should have the ability to stand still at a bus stop with closed eyes and have the entire universe disappear apart from a single blazingly visualised sigil or muttered spell.

Drugs have a long history of use in magic in various cultures, and usually in the context of either ecstatic communal rituals or in personal vision quests. However compared to people in simple pastoral tribal situations most people in developed countries now live in a perpetual state of mental hyperactivity with overactive imaginations anyway, so throwing drugs in on top of this usually just leads to confusion and a further loss of focus.

Plus as the real Shamans say, if you really do succeed in opening a door with a drug it will thereafter open at will and most such substances give all they will ever give on the first attempt.

Chapter 8
The Spell of the Narration

$$U^2 = H = \Phi + S/k$$

This Octavo (Roundworld Edition) provisionally had the subtitle of:

'Low and High Magic in a spatially and temporally finite and unbounded eight dimensional hyperspherically vorticitating universe with a wave-particle duality, quantum indeterminacy, and a high value of lightspeed relative to the size of most organisms wishing to practice it.'

However that wouldn't all fit nicely on the front cover so we compromised with:

'A Sorcerer-Scientist's Grimoire.'

And now we approach the High Magic bit........

The Map part 4.
The conventional Flat Universe cosmology story includes an astrophysics sub-plot that suggests we inhabit a universe that runs down from some inexplicable condition of very low entropy at the big bang towards a terminal state of complete entropy with all the stars gone out or swallowed up in black holes. This sad ending to the astrophysics story depends on two assumptions.

The first depends on the idea that the lowest and the highest energy matter particles which have stability generally lack antiparticles in this universe. Thus matter has a form in which it can get permanently stuck.

Secondly it critically depends on the idea that entropy increases as the universe expands.

The first assumption about the stability of matter depends on the conjecture that neutrinos (the lightest matter particles) and neutrons (the heaviest of the stable* matter particles) both behave as Dirac fermions which means that they need separate antiparticles to annihilate against.

(*neutrons at least appear fairly stable in bulk or in atoms although they tend to decay rather quickly when they fly free).

Now evidence mounts that neutrinos may in fact act as their own anti-particles, that they behave as Marjorana rather than as Dirac fermions and this throws into question the entire meaning of anti-matter and its apparent almost complete absence from this universe.

The HD8 hypothesis advanced in The Apophenion suggests that neutrons will also exhibit Marjorana fermion behavior under sufficiently extreme conditions, namely in the raging hearts of neutron stars. In this case matter does not have any stable default states into which it degenerates. In heavy old stars the Dirac type electrons and protons collapse together to form neutrons and these will then annihilate against each other liberating blistering amounts of energy which prevents further collapse of the dying stars into black holes or singularities. As the energy particles fly back into space away from the vast compressive forces the reaction reverses repopulating empty space with free neutrons which decay back to electrons and protons which eventually combine in cooler conditions to replenish space with its thin haze of hydrogen.
We all know about hydrogen, it consists of a colourless and odourless gas that gradually turns into people.

Hydrogen, the simplest of all the elements consists of a single electron orbiting a proton. Of all substances this most resembles the alchemical prima material for stars can transmute it eventually into any other element.

Billions of years have to pass just for some of it to coalesce into medium sized stars. Then you have to wait a while longer whilst those stars get round to serious nucleosynthesis [25] and making heavier elements before finally exploding. More time passes and eventually a new star forms in the vicinity and sweeps up all the heavy element debris from the previous one to make orbiting planets. If all that goes well and the cooking temperatures remain within reasonable limits, then micro-organisms develop and within a mere blink of cosmic time some of those bacteria evolve into us.

As we comprehensively argued in Chapters 1 to 3, this universe does not expand despite the optical illusion to the contrary. It has the same size at any point of the finite and unbounded time within it. So, as argued in Chapter 4 the universe always contains the same number of bits of information as defined by its surface area. The size of the book remains the same but it constantly rewrites itself as the entropy and the information shift around and mass turns into energy and back again with a restless creativity and destructiveness.

The relationship between information and entropy has a far greater subtlety than a simple equivalence as we shall see.

We have incarnated within a never ending story that has no cosmic scale genesis and no impending cosmic apocalypse, despite all the small to medium scale genesis and apocalypse going on all around us. Quantum indeterminacy confers upon us the choice of what sort of chapters we would like to write for ourselves. Here we come at last to what you might call 'Spiritual Matters'. What makes a good life story, and in particular what should we use our freedom and magic for? What does indeed constitute 'High Magic?'

Distilling metaphysics from physics and other sciences and then further distilling those metaphysics to obtain ethics always remains a difficult and dangerous alchemical exercise with occasional ghastly consequences. Newtonian dynamics conceivably leads to an atomized sort of perspective and hence to a hypertrophied individualism and an exploitative approach to the natural world. The horrors of the French revolution and Bonapartism derive partly from taking Enlightenment ideas too far.

Darwinian theories seem to have led to some ill-conceived eugenic practices and also fed into fascist and capitalist theories, whilst neo-Lysenkoist biological theories underpinned much of communist ideology.

The newer relativistic and quantum paradigms and the science of ecology seem to point towards a more holistic metaphysic of interdependence between all aspects of the cosmos and between humanity and its environment, and perhaps towards a mystical panpsychism. Yet this in itself does not seem to yield any particular system of values and ethics.

However if we look at what the universe actually does, or rather what we think it does, then we can at least make a stab at aligning or own value systems with what we think the universe strives for.

The universe cycles its parts through a ceaseless dance between order and entropy, concentrating energy and then dispersing it again, yet wherever energy flows, complexity flowers to hijack the local flow towards entropy, and matter self-organizes itself to make galaxies, stars, planets, chemistry, and then almost certainly biology, bacteria, bugs and Beethoven or something similar. So if we ascribe a positive value to existence then we tacitly ascribe value to complexity. We should perhaps do so more explicitly and thus justify the intuition that values a finely wrought object or idea or item of art over a piece of junk, a sophisticated organism over a less evolved one, and a wise person over an idiot, even though the existence of the former in each case often depends on the existence of the latter. Yet in each case the intrinsic value lies in the amount of coherent or integrated information that a structure possesses, and to hell with political correctness, cultural relativism, and rubbish art and architecture.

Chaometry part 5: 'Spiritual' Matters

The Eighth spell/ equation shows the relationships between boundaries, information, 'negentropy' and entropy for any system and in doing so it defines the size and limits of any story that the system can create.

In the form shown: $U^2 = H = \Phi + S/k$, it relates to the entire universe, however we can easily cut it down to planetary ecological, socio-political, or to personal size by adjusting the boundary parameter from U, the Ubiquity constant, to something smaller.

It works for any closed system where we can specify the boundaries beyond which information does not pass, or an open system through which a measurable amount of information passes, and so it can model the rise and fall of societies, empires, civilisations, religions, ecosystems, and even individual human lives.

Discworld contains a number of abstract concepts and philosophical ideas in an actual physical or anthropomorphic form. 'Death' for example exists as only a privative concept in Roundworld, it merely denotes the absence of life. Death stalks Discworld as an actual character engaging the about to become departed IN MORDANT AND PORTENTIOUS CAPITALISED CONVERSATION before wielding his scythe.

Discworld also contains an extra physical chemical element called Narrativium which constrains stories to make satisfying sense despite the lower levels of scientific causality and the higher levels of magic which pertain there.

Here on Roundworld we have an analogue of Narrativium which takes the form of abstract concepts linked by an equation, the so-called Spell of the Narration, or the Eighth Equation. Whilst this does not necessarily guarantee satisfying stories or happy endings it does allow us to make some general sense of, and plans for, stories great and small, despite the huge element of randomness here on Roundworld.

The Eighth equation has the general form:

$$[U^2] = H = \Phi + S/k$$

Ubiquity Squared, $[U^2]$ shows the effective 'surface area' of the system in units corresponding to information. For the universe itself Ubiquity Squared yields the simple answer of $\sim 10^{120}$ bits. For messier systems like ecological ones the surface area usually corresponds to something like the total possible photosynthetic surface available.

For individual humans the 'surface area' effectively consists of their social, geographical, financial, and mental boundaries.

H means the information available in the system under consideration.

S means the entropy, the useless unstructured unreadable information that at its lowest level just consists of random molecular motions. We need to divide S by k, Boltzman's constant to represent it in the same terms as information. Basically as Boltzman's constant has a very tiny value, a small amount of entropy equates to a huge amount of information.

Oddly enough the unit of entropy does not have a proper name, just the symbol S and the units of joules per Kelvin. We suggest calling it the Bolt after poor old Ludwig Boltzmann who discovered entropy. Despite having a world class beard and ideas to match, he later chose suicide.

1 Bolt of entropy would thus equal roughly 10^{23} bits of information.

Φ, Phi, equals the 'Integrated Information', after the concept of Tononi[26].

It remains rather tricky to define or quantify but it basically implies 'negentropy' or the amount of structured information exchange possible between the elements of a system, and the degree of sophistication of the elements of the system.

So now we see that the information H, available to a system appears as $\Phi + S/k$. If the system has little or no structure then it all appears as entropy, if the structure and organisation goes up then the entropy goes down. If the available information locally increases or decreases then one or the other of Φ or S/k or both, have to change. These sorts of events cause the rise and fall of mice and men and entire star systems.

The Eighth equation finally ensnares entropy within Noether's Principle. Of all physical quantities it alone appeared to violate any conceivable conservation law by apparently increasing inexorably. The Eighth equation however shows that the surface area of a system conserves the information inside and that the total entropy and integrated information within it remains constant.

The rise and fall of the Roman Empire provides a classical demonstration of the equation in action. The empire had only a single source of energy in the form of photosynthesis by the plants which produced its human and animal food and its wood. The geographical limits of the Empire thus defined its energy availability.

Now the sunlight energy captured by plants basically acts as an information supply because it underlies absolutely everything that an agrarian based system can do from increasing its population to establishing armies and bureaucracies and to building cities and writing poetry.

The Romans developed many highly effective and highly complex forms of Integrated Information. They spent their energy/information input building a tremendously complicated agricultural, social, cultural, religious, military and administrative structure.

Many of these structures served to reduce entropy, to reduce the amount of incoming sunlight wasted on unfarmed land; hence their farms became huge and extended into previously under-farmed territory. Of course this required ever more complex systems of Integrated Information to make it work. It also supplied an impetus for the expansion of the Empire by conquest.

Eventually the whole system entered the realm of diminishing returns where it had to spend ever more of its incoming energy/information on maintaining an ever more complicated and far-flung system.

Critically, the Romans had failed to make any significant technological or organisational innovations for centuries; they just relied on making the system bigger rather than more sophisticated. Even a half sensible arithmetic system might have helped. Instead of improving the quality of their Integrated Information they just increased its quantity.
The system just got bigger without getting any smarter.

After passing the diminishing returns point the Empire found itself increasingly vulnerable to challenges from such events as poor harvests, disease, and barbarian incursions which it had overcome in the past.

The Empire fell much more quickly than it had arisen. The high degree of interdependence between all the elements of the Integrated Information system meant that a failure in one part tended to bring down all the other parts. The Empire underwent a radical simplification as the Integrated Information structure broke down and the dark ages began. Only one part of the empire's structure survived; a hastily cobbled together and radically simplified state religion. Christianity did not cause the fall of the Roman Empire. Christianity arose as a symptom of the general collapse of its complexity.

The Empire could no longer afford to sustain a rich and diverse religious culture so it downgraded to a much simpler model which allowed for tighter social control at the expense of a freedom of thought which might possibly have saved it. An explosion of religious diversity has sometimes occurred when a political system gets in a mess.

Entropy then increased all over the former territories of the empire, land reverted to the wild and the human population crashed taking with it a sophisticated culture.

The ghastly murder of the worlds leading female scientist, Hypatia of Alexandria and the tragic destruction of the worlds learning in the library of Alexandria by a demented Christian mob sent by the new Christian bishop Cyril, stands emblematic as the start of the dark ages.

Our own civilisation now faces a similar scenario. We have built a fantastically complex and interdependent system of Integrated Information based critically on the energy/information input from fossil fuels. However we have now entered the realm of diminishing returns from this resource. We have already used vast amounts of it merely to create entropy and huge amounts of low grade throwaway Integrated Information instead of quality creations. If we cannot innovate and substitute some other inputs such as nuclear or renewable power then a systemic collapse will follow and most of our population and culture will go with it. End of story.

On an individual level we face the same choices as entire civilisations. We can look for new horizons and boundaries to change our energy/information input. We can use the input to increase our Integrated Information either in quantity or quality, or we can just squander it away on entropy.

Magical Ritual part 5.

So now we come at last to the part which should perhaps have come first: -

The Choice of What To Use Magic For.

An inner logic prevails here, hopefully the aspiring magician will have read this far and will heed these words before action, for here we approach the big questions, what some have called the 'spiritual' questions, how to live a fulfilling life. It now seems possible in the light of the Eighth equation, to apply a certain metaphysical calculus to this question quite independently of all the somewhat ad-hoc moral and religious philosophies.

When composing a life story, either with magic or without, four meta-choices apply:

1) Change Horizon Size? $[U^2] = H$
2) Increase Integrated Information Quality? Φ
3) Increase Integrated Information Quantity? Φ
4) Increase Entropy? S/k

1) Individual's horizons consist of all that they know and all that they have experienced plus all that they can do. Such horizons define their energy/information input.

Broader horizons beckon in youth; nearly everyone wants to increase their experiences and capabilities and some pursue this quite recklessly and actually end up limiting their horizons. Wealth, travel, socialisation, reproduction, new experiences and learning generally increase an individual's horizons but each can easily prove a limitation if its pursuit narrows in specialisation and if it starts to give diminishing returns or if it begins to exclude all else.

Within the chaos magic paradigm the magician can perform results magic or invoke a deity (a real or imaginary internal power source) in support of virtually any activity in life.

But, in general, only those conjurations likely to expand the horizon size seem worth performing.

Do, or conjure for, only such things that will probably lead to greater opportunities or abilities or experiences or knowledge. Hubris informs magic at its best. Although we shall probably die trying, we seek the wisdom and the power of whatever gods we can imagine in this Promethean quest.

2)/3). Integrated Information costs the same to produce in terms of effort whether it consists of high or low quality information. Producing an illuminated bible with exquisite gold leaf encrusted illustrations and a hand copied text consumed the best part of a human lifetime, so did the creation of the theories of Special and General Relativity. A familiarity with the complete works of Aleister Crowley costs no more effort to acquire than a familiarity with all the scores and names in some sport or other for the last 20 years.

Quality costs no more than Quantity when it comes to coherent information, much as fitness costs no more than fatness in terms of the effort required achieving it. Fat people merely carry around a lot of surplus useless information that fit people have used doing something more interesting. Plus they have less robustness and fewer options. And so it goes with the mind and with entire cultural paradigms. More of the same doesn't count as better unless you haven't enough. Different counts as better, particularly if it might lead to something else.

In first world nations we inhabit very fat cultures that have come to rely on the production and consumption of excessive quantities of mediocre products and ideas and experiences. Such cultures have thus undermined their own robustness and adaptability and have rendered the identities of many of their denizens rather fragile and hollow.

Industrialised cultures have blown the energy/information stored in the planets fossil fuel reserves on over-breeding by reducing the selection pressure on unpromising stock and then failing to even educate most of the progeny to anything like their capacity. Quantity seems to have triumphed over quality even in the production of people.

Consumer choices now define identity choices for many people. The magician finds more satisfaction in creation than acquisition, no matter what others think of the creation.

The integration of a new piece of mental furniture into ones world view, the creation of a piece of art, the mastering of a new ability, the forging of a new relationship or friendship, all of these bring so much more lasting satisfaction than a mere purchase.

Conjure then, for needs if must, but preferably for opportunity and quality of experience, but never for merely more of the same.

Conjure not for wealth, but for the experiences that you would spend the wealth on if you had it. Any necessary wealth will then materialise as a side effect.

4) Entropy. Several thousand years ago the ancient Greeks realised that Virtue trumps Pleasure when it comes to life satisfaction. That realisation remains widely ignored to this day.

Any human activity with the exception of sustainable agriculture, reproduction, and thinking creates more entropy than useful information.

Virtue accrues through increasing the amount of quality Integrated Information within your horizon through self-improvement, and service to others. You can never loose virtue (although you can make an equal or greater quantity of anti-virtue as well by self destructive or selfish behaviour).

As homeostatic organisms we remain incapable of making any long term change to our overall pleasure levels, except in the case of severe privation. So it hardly seems worth pursuing except as a side effect of something more interesting.

Pleasure does not accrue at all; rather it proves all too ephemeral and evaporates quickly leaving no lasting satisfaction or useful information. Rather then we should seek forms of virtue that produce pleasure for us as a side effect to balance the sometimes irritating and painful aspects of the quest for virtue.

The Great Work of Magic, rather like The Great Work of Science or The Great Work of Art, lies in creating more high quality Integrated Information than entropy.

It should create ideas that will stand the test of time or lead to new ways of thinking.

Though frequently despised and persecuted, magicians and the magical style of thinking have led both directly and indirectly to the creation of several technologies and sciences including metallurgy, chemistry, astronomy and medicine, the creation and destruction of entire religions and the evolution of numerous art forms.

The particular special skill of the Magician Caste lies in its willingness to investigate what others consider impossible, insane, inviolate, and unthinkable.

Magicians continue the practise of Natural Philosophy – the investigation of the workings of nature. The term 'Natural Philosophy' now often implies the mere precursor paradigm to modern scientific ideas, but it does embody the perpetually revolutionary idea that even seemingly miraculous events lie open to understanding rather than consigned to mystery.

Magicians in all eras have tended to use the magical perspective as a jumping off point and an inspiration to study a diverse suite of disciplines. At the time of writing, the fields of psychology, physics, nanotechnology, comparative religion and ethnography, and ecology seem to attract their particular attention.

Magicians consider that modern science delineates itself somewhat too tightly and often misses the bigger picture, the anomalies that do not fit the system, and the events for which it can find no mechanism.

Many magicians remain content to use the protective cover of Natural Philosopher to describe their interests, as many of their colleagues did in centuries past.

If we ever get to the stars we will get there using tricks that will come from insights that derive mainly from what we now call magic rather than science.

A little results magic in support of what you need in life remains acceptable, so long as it remains in the service of the Great Work of High Magic, the quest for greater Knowledge and Ability*.

*(Note the studied avoidance of the word Power, which has some rather worrisome connotations at our current rather primitive developmental level, just 10,000 years out of the caves).

Please conjure responsibly.

Appendix 1. Chapter ?
The Knights of Chaos

An Appendix specific to the 21st Century, wherein we speak of Eschaton, Apocalypse, and The Knights of Chaos.

Chaos always has the last laugh.

*(Despite that **Death** and **Entropy** both rather tiresomely insist on claiming this honour. Life beats death in the long run despite and because of, individual deaths. Inexorably increasing entropy applies only to relatively small closed systems.)*

The Orthodox and Canonical Roundworld Octavo Grimoire, cannot possibly have more than Eight chapters and we do not particularly care for the number nine anyway, so we adumbrate this as Appendix 1, or Chapter (?) instead, as it resembles a 9, and also signifies the major question facing our species at this time, do we want an apocalypse or not, and if so, of what kind?

Chaos always raises more questions than answers, we will always have chaos but do we want the high grade evolutionary chaos or the low grade devolutionary chaos?
The Discworld Planck's constant has a pretty high value and although it has a rather smaller Ubiquity constant, the value of $\hbar_D \sqrt[3]{U_D}$ nevertheless comes out at about 50 times its Roundworld equivalent. This makes Panpsychism and Hylozoism much more easily observable on Discworld and it leads to a ready anthropomorphic personification of things that usually appear only as concepts on Roundworld. On Discworld, concepts walk around with scythes and swords and struggle autonomously to implement their own

agendas. Here on Roundworld we have to carry them round ourselves and give them a bit of extra belief and assistance.

The magical field of Discworld readily allows Chaos to appear as an actual character with useful godlike attributes. Roundworld magicians, particularly Chaos Magicians, may wish to consider investing the extra effort required to give such a concept an actual godform here, the better to invoke its interesting powers.

The Discworld mythos of the character of Chaos exhibits a considerable metaphysical sophistication which rather miraculously seems to parallel the development of the concept of Chaos on Roundworld.

Chaos started out as Khaos (as our Greeks know), a sort of unpredictable primordial mess from which the ordered universe developed. On Discworld, Khaos has a half remembered past as the possible origin of all form but he then functioned as the Fifth Horseman of an Apocalypse, a time when everything would revert to a sort of unpredictable primordial mess with the help of Death, War, Famine, and Pestilence. However on Discworld they call it an 'Apocralypse', implying that it all sounds a bit apocryphal and possibly avoidable and that it doesn't have to happen.
Similarly we don't have to have a cosmic apocalypse here in the Roundworld universe if the Big Bang didn't actually happen, and we can avoid a homespun ecological catastrophe if we choose to.

For a while Khaos gave up the apocalyptic role, resigned from the Horsemen, adopted a rather more symbiotic relationship with order, and took a name change to Chaos. As the world got more civilised and vastly more complex and law-ridden, so its unpredictability increased as did its subjection to the sort of 'butterfly' effects as investigated by chaos mathematicians. Whilst the old Khaos ran on a complete absence of rules, order and structure, the new Chaos runs on the chaos that arises out of complexity itself. Thus instead of the initial Khaos versus Order duality, a more subtle symbiotic triality of Order-Chaos-Disorder emerged.

Eventually Chaos rallied the horsemen and averted apocalypse on Discworld.

However Chaos maintains a healthy disrespect for Order and The Law, subverting it for a living and going to war against it when eventually riled. Despite the name and role change Chaos has nevertheless retained a classic style for formal occasions, and the Discworld avatar manifests as follows:

Chaos appears concealed by a full face helmet with a design somewhere between a butterfly and a Rorschach blot covering the visor and wielding a weapon so cold that it has negative heat. The weapon symbolises **the power of Chaos to reverse entropy and to bend the laws of probability**.

These sound like exactly the sort of general abilities the magician needs to focus on, so it seems like a form worth investing some effort in.

Now the Roundworld concept of Khaos started with the classical Greek idea of a ghastly, unpredictable and unlawful mess from which emerged first the world, and then the rather nasty Titans, and then the slightly more agreeable gods and eventually humanity, with the implication that it could all go Khaos-shaped again if we don't behave.

Subsequently chaos just became a synonym for disorder, meaningless randomness, anarchy, system failure, and destruction. However a few contrarian souls could still perceive the positive potential embodied in the concept of chaos, creative destruction, adventitious change, and unlimited possibility.

One must have chaos within oneself, to give birth to a dancing star. ~ Friedrich Nietzsche

In this spirit the chaos magicians of the late twentieth century began testing the rules of reality to breaking point, and then some........

Under the banner of 'Nothing is True', Everything is Permitted*' they take nothing for granted from conventional magical theory and practice, or from science, theology, psychology or parapsychology.

(*Later amended to 'Nothing has Absolute Truth, Anything can Happen', on the grounds that any use of 'is' or 'being' represents wishful thinking rather than Illuminati grade thought.)

Around this time Roundworld, mathematicians discovered that many processes that seemed random actually showed extreme sensitivity to initial conditions. Thus we now understand why we cannot solve the three body gravitation problem, and why we cannot predict next month's weather and a vast number of other moderately complex things. In fact it seems that we can now prove that we cannot really anticipate what any complex system will do beyond a certain point. To know such things we would require impossibly precise data about the starting conditions because in many systems a minute difference at the start rapidly mushrooms into a completely different future. Yet this will happen to ANY system eventually.

So Chaos Science, Stories from Discworld, and Chaos Magic all appeared on Roundworld at about the same time, and now some decades down the line, we attempt to draw them all a little closer together.

The new Chaos Science began to use the rather appalling term 'Deterministic Chaos' to describe the behavior of the systems that it investigated. The underlying assumption remained that the system didn't actually behave randomly but with such sensitivity to initial conditions as to render precise prediction impossible. Thus it didn't offend the causal sensibilities of classical scientists, but rather it seemed to reassure them that even seemingly unpredictable events still ran on cause and effect principles rather than on truly indeterminate randomness, or proper Chaos.

However Chaos Science seems to have missed out on the Quantum perspective that informs a good deal of Chaos Magic and the Discworld stories.

Extreme sensitivity to initial conditions has no theoretical limit in chaos science. Thus it extends down to the quantum domain underlying physical reality where events become indeterminate. So it acts as a sort of ladder by which quantum randomness climbs up into the macroscopic world to eventually create huge indeterminacies. So not only might a butterfly changing direction over Mongolia eventually trigger a hurricane in Florida, but a random quantum quirk in the butterfly's nervous system might have contributed to the direction change in the first place.

Thus massively improbable things happen all the time. All events that occur partake of this fantastic improbability. At the times of our conceptions the chances of us encountering each other on precisely this page would have stood at almost infinity to one against.

The whole universe runs on a Total Improbability Drive, Chaos Rules!

The Knights of Chaos

Averting Apocalypse

The gods come from Chaos as our Hellenic ancestors surmised. The deeper you penetrate into Chaos the more primordial, powerful, and unpredictable the deities you encounter.

The Elder Gods of Sex and Death, of Love and Hate, and of Fear and Desire, lurk beneath the thin and fragile veneer of our civilisations.

Yet beneath this pantheon lie the Eldest Gods of all, the raw forces of creativity and destruction of the universe itself.

The universe has a Panpsychic Sentience of a sort. It lives. Its components exhibit random quantum autonomy and they communicate with each other non-locally to produce a spectrum of 'consciousnesses' from the sub-atomic through the biological to the galactic.

Most of these 'consciousnesses' appear so alien to us humans that we can barely recognise them as such. Fellow mammals seem partly comprehensible, reptiles seem less so, insects remain inscrutable, and as for atoms and galaxies, well we merely have a few flimsy and paradoxical equations with which to understand them.

Our appreciation of Chaos has evolved from a primitive dualistic view of Order versus Chaos into a more sophisticated triality of Order-Chaos-Disorder in which Chaos not only undoes things but it also acts creatively to ferment and evolve new Order and fresh complexity.

Magicians seeking to invoke Chaos need to carefully consider the cost - benefit analysis of such Faustian Pacts, for Chaos can manifest on a spectrum from random destruction to random creativity. Making anything always involves breaking something else; the magician should consider carefully the size and quality of any omelette that may possibly result.

The magical technique of ADP, Anthropomorphic Deity Personification, remains the magician's most effective method of interacting with events that we can at best in rational mode, usually only conceptualise as abstract concepts.

Chaos phenomenises in fractal mode, from quantum particle level, through its chemical, biological, ecological, sociological, psychological, technological, and political manifestations, and then on up into astronomical and cosmological levels.

At the most basic nuclear level such anthropomorphic personifications as H.P. Lovecraft's conception of AZATHOTH, the 'blind mad god at the centre of chaos' allow the magician to channel pure randomness to macroscopic levels.

Of course unadulterated randomness will most probably have deleterious effects on the health, lifestyle, and sanity of relatively highly structured organisms such as ourselves, and few take such risks, except in extremis. Magic may well fill the vacuum when you have exhausted the resources of common sense, but conjuring for a purely random result implies either a great faith in the universe or perhaps a great faith in one's own deep subconscious.

Sometimes it works, sometimes it doesn't. For every successful artist, hero, and mystic, hundreds lie dead and forgotten, their random inspirations led to nothing more.

At the more complex levels of reality the interplay between Chaos, Order, and Disorder becomes more sophisticated. Order strives to control Disorder but it also provides new possibilities for Chaos, and Chaos spawns both new Order and Disorder.

The more highly ordered a structure gets, the greater the likelihood of catastrophic failure, unless it retains enough chaos to restructure itself.

As we enter the somewhat arbitrarily designated third millennium the Order of the human world seems perilously perched on the brink of Apocalypse.

Stand on any hillside overlooking one of our huge cities and notice its fundamental instability. Any interruption to the endless food convoys streaming into it or to the megawatts of power and fuel flowing in spells disaster, as does breakdown in its internal communications media. Modern cities lack the robustness of pre-modern towns which could usually supply themselves from the surrounding countryside, and restore command and control functions with a short walk and a talk.

Our numbers and our consumption have reached levels that our technology cannot sustain for much longer.

We have severely depleted our planet's fossil fuels to achieve such levels of population and consumption.
We face catastrophic climatic and ecological changes if we continue on our present course.

We do not yet appear to have many viable alternatives to this Apocalypse on the horizon.

The world's richest nation now needs 10,000 gallons of oil to survive each and every single second. It has few plans to reduce that dependency, and most other nations now seek to emulate this behaviour.

We have already wrecked large areas of the planet's oceans and our new unsustainable industrial farming techniques imperil the fertility of the land.

On current projections the human population of earth will climb to 9 billion by mid-century. The end of century figure remains debatable but it may not exceed a few hundred thousand living in Medieval or perhaps even Neolithic barbarity in the Polar Regions only.

The four horsemen, Famine, Pestilence, War, and Death, groom their steeds. Within a few decades most of our cities and civilisation may lie in ruins as the food and the fuel run out and we slaughter each other over what little remains.

This has happened many times before on local scales, it may soon happen on a global scale.

Any fall tends to occur much more quickly than the long and apparently comfortable climb up the slope to the cusp of catastrophe.

Unless of course we evolve other plans.......

Whether we act or not, we seem to face an impending Eschaton, a global change of planetary ecology and climate and human civilisation. A slim chance exists that we can Immanentise an agreeable Eschaton rather than a vast historical reversal. It seems worth a try.

Cue the entry of the Fifth Horseperson, Chaos, who might just show us how to distort probability and arrest the slide into entropy sufficiently to create a merry Apocalypse instead, in which the impending Apocalypse becomes but a mere myth which we had the sense and inspiration to avert.

An Invocation of The Knight of Chaos.

-1) Prologue. Chaos or Kaos appears in a number of ancient cosmogonies as a gaping void or abyss from which creation arises, and it also finds representation as various chaos creatures such as Typhon, Taimat, Leviathan, and also in the form of Dragons, all of which engaged in wars with various culture heroes who try to establish order.

The Greek personification of Kaos has either an indeterminate gender (that figures) or a female one (so does that). However Discworld has a male Chaos figure that has a relatively mundane day job and name when off duty, as do many of us.

Mathematical Chaos of course has no gender, and neither does the chaotic creative function of the universe itself. Thus in this chaos magic style invocation of Chaos the godform has no particular gender, so magicians of either, neither, or both persuasions may attempt it.

The Chaos godform in this invocation has a particular use when the magician wants to conjure for something highly improbable and thus enchants long by casting a spell to affect events well into the future. It takes the name of The Knight of Chaos. It represents not the whole spectrum of Chaos but rather the questing aspect of it, the OBDAXAZONGAGA, the 'higher' chaos of complexity evolution. In mythos terms we can regard The Knight as the child of Ouranos and Apophenia; high magic coupled with inspiration. By invoking The Knight of Chaos the magician assumes the identity of A Knight of Chaos in the quest to Avert Apocalypse

0) Preparations.
Commit the Vernacular invocation and the Ouranian Barbaric Charge and Incantation to heart.

Fashion a mask or full face helmet that incorporates a Rorschach blot in a shape suggestive of a butterfly.

Prepare a large magical instrument suggestive of a sword or a wand or something of both.

Prepare also a small concealable magical sidearm.

Prepare an Octaris, an eight rayed chaos-star for visualisation, and also a Knight's Star, either in the mind's eye or in physical representation.

Select incense if desired, Bitumen or Asphalt serve well for dark works of attack, Ouranian blends of Valerian Root and Oakmoss serve well for most other purposes.

Prepare any necessary materials and ideas for any intended conjuration, which may consist of works of Enchantment, Divination Evocation, or Illumination.

1) The Proclamation/Statement of Intent.
'I/We do will to invoke the Knight of Chaos, and to launch this conjuration for.......'(word as appropriate.)

Intone the Ouranian-Barbaric charge: -

CHACUCHAD	VONZOG	CHIPUVT VIVASSAGITSIRIF
Nothing	True	Everything Permissible

2) The Centring and Encircling.
(By any preferred method to taste.)

3) The Negative Gnosis.

Visualise or gaze upon the Octaris, the eight rayed star of Chaos for eight or a multiple of eight very long slow deep breaths.
Then strive for a few moments of complete mental voidness.

4) Invocation.
Take up the mask and weapon.
Trace the Knight's Star in the air and the aether with the weapon.
Deliver the Vernacular Invocation.

5) The Positive Gnosis.
Commence hyperventilation and/or whirling or other chosen excitatory gnosis, with further visualisation or gazing on the Knight's Star.

6) Incantation.
Deliver the Ouranian Barbaric Incantation. Perform glossolalia if moved to do so.

7) Conjuration for desired effect.

8) Laughter Banishing.

Vernacular Invocation.
First you knew me as Kha-os,
Primordial Void that gives birth to desire,
To the World, to the Titans,
To Darkness, Hell, and Old Night.

So then come the Chaos Wars,
They call me Typhon and Dragon,
Leviathan, Taimat, and Devil,
Mother of hell-spawn that gnaw at your law.

Your golden heroes do not prevail,

System and order eventually fail,
Chaos remains for ever,
Author of heaven and hell,

Mathematical entrapment I abhor,
The calculators they would imprison me,
With numerous tricks and glamorous pics,
But I get out of that with a quantum fix.

At realities roots, I make probability,
I gestate and pupate it,
My butterflies mutate it,
Random disorder and bugs in the system!

I come masked in confusion,
To undo delusion
With chaotic profusion
I make new illusion

No darkness and hell
No fear of disorder
No number confines
My dancing star shines

With Improbable sword
That stirs up the void
And selects from the mire
I phenomenise desire!

Ouranian Barbaric Invocation.

GEHURKAL
The Great Old One

UFOSETH
Fear

HAJAK
Kill

SUNDEGAI
Destruction

EROX
Chaos Star

WIXAHAFIS
The Authorities Arrive

CUNGEVAAB
Do The Great Work (of)

AEPALIZAGE
Immanentise the Eschaton

HADAKA
Serpent

KULHAL
Dragon

REHOHUR
Destroy

BAGUNGA
Fighting

HODOJEPA
Chaos Magic

TZONGA
(in the)
Unpredictable

LEVIFITH
Magic

AHIKAYOWFA
(of) Higher awareness

UDINBAK
(of) Chaos

KUDEX
(of) Darkness

CHENCEB
Consume

XEBEMEK
Dividing

TABOCH
Yes

SXAUL
Temple of Regularity.

OBDAXAZONGAGA
(of)   Higher Chaos

CHOYOFAQUE!
Do The Great Work

The Knights of Chaos.
The chances of advanced human civilisation surviving the twenty first century now seem to diminish with every passing year. As a species we stand in dire need of creative shocks and surprising innovations to our ways of thinking

In some ways Magicians got humanity into its current impasse by stimulating scientific enquiry in the first place but this has led to unsustainable forms of hyper-industrialisation. Now we must search for new and alternative futures. Science on its own now seems alarmingly short of technical fixes for the problems that its technology has spawned.

The Illuminati now summon The Knights of Chaos in response to this situation.

Occasionally The Knights will need to conjure for minor apocalypses by arrangement to frighten humanity away from the global meltdown of civilisation.

Occasionally The Knights will need to conjure for and against various memes and those who play host to them.

Occasionally the Knights will need to debate the case for conjuring for what now seems impossible.

The Knight's Star consists of the following symbol that we use for mutual recognition: -

Figure 8. Knight's Star

This star of five lines and eight points symbolises the Magical and Chaoist affinities of The Knights and their quest to Avert Apocalypse.

The Knights consist of those capable and willing to phenomenise events and ideas of absurd improbability in the cause of The Aversion of Apocalypse.

Some have this erratic ability naturally; some can learn to occasionally manifest it. The requirement exists for a substantial host in place by the end of the first quartile of this forbidding century.

The absurdist title of The Knights of Chaos provides a convenient cultural camouflage. Any number of fantasy-gaming and role playing societies now use similar honorifics, having borrowed extensively, for purely entertainment purposes, from a mythos that Chaos Magic helped to establish. This can prove useful if the forces of awe and boredom take an interest in the presence of strange paraphernalia and rather dangerous looking instruments.

We just belong to a fantasy RPG group, the thought of an occult conspiracy to overthrow consensus reality, to confound popes and princes and politicians, and to reset the course of human history never entered our minds for one moment Officer.

The quietly fearsome Sororres of this chivalric magical order rarely title themselves Dame, nor do Fraters call themselves Sir in public. Rather all tend to use the epithet K.C. in appropriate private correspondence.

Recognition as a Knight of Chaos may occur following the fulfilment of certain conditions and various unspecified uncertain conditions.

Applicants should learn the Invocation of the Knight of Chaos ritual.

Applicants should equip themselves with a suitable black robe, helmet or mask, a large ceremonial magical instrument, preferably of Pole-arm size, and a smaller concealable magical sidearm.

Applicants should provide evidence of at least some miracles of probability distortion and a suitably humble (humorous, non-egocentric, or self-deprecatory) attitude towards them.
As the phenomenon of personal existence has such a vast improbability anyway, we should not get too egocentric about occasional acts of parapsychology.

Applicants should agree to remain on short notice call by some appropriate medium for conclave or magical response, depending on global circumstances.

The position of Marshall of the Knights of Chaos shall change by general vote as circumstances develop. At the time of writing Specularium and Arcanorium provide points of e-contact.

Chaos Magic helped to change the way that a substantial proportion of the world's magicians think about HOW to perform magic. Hopefully this initiative will add further considerations about WHAT to use magic for

Figure 9 The Knight's Seal

Appendix 2
The Mass of Chaos E, to ERIS

Prologue:

The modern conception of Eris seems to have begun with the publication of the counter-culture Principia Discordia which built upon the classical Greek mythos of the troublemaking sister of Ares the war god.

Whether Discordianism represents a joke masquerading as a religion or a religion masquerading as a joke remains undecided, but she soon found favour with the newly assertive feminine faction and its sympathisers.

Her mythos bears some resonances with the Hebraic myth of Eve who mightily annoyed the grim old patriarchal god JHVH by taking the apple from the tree of knowledge.

Magicians of both sexes have found in her a lot of just what they wanted and needed in a goddess, and often a lot more besides....

The following ritual invocation developed from work begun in 1990 and has remained in constant use with many variants and adaptations since.

The incantation appears ensigilised in the Enochian magical language as this invocation predates the use of Ouranian Barbaric. Certain lapses in the use of Vernacular Prime appear corrected in this version below.

Commentary:

Legend has it that the Greek Goddess Eris hurled down a golden apple marked KALLISTI (for the fairest) Goddesses to create a riot, the reverberations of which launched the thousand ships. As a Goddess of Chaos she appears as an archetypal muse who brings inspiration and creative confusion through the clash of opposites. Whenever an irresistible force collides with an unmovable object, whenever woman collides with man, whenever we find paradox, when yin meets yang, in that moment of fertile confusion, pregnant with new potential, we detect the hand of Eris stirring things up out of pure wanton curiosity. In the manner of a crazed chemist she playfully mixes her most powerful reagents to see what explosions result. Heedless of the entropy in her wake, she delights ever in the wildness and weirdness she provokes.

We traditionally invoke Eris in the form of a goddess with flaming red hair, dressed in a riot of colour and carrying an apple marked K for Kallisti and Kaos. Her personal sigil takes this form, known inexplicably as the five fingered hand of Eris:

Figure 10. Eris Symbol

This represents the waxing and the waning moon, the horns of the shamanic life force spirit and also two arrows meeting in a head-on collision.

She resumes the qualities of the Babylonian Goddess Ishtar of Love and War who took an eight-rayed star as her emblem, Kali the Hindu Goddess of Sex and Death, and by Babalon and Lilith, the Hebraic conceptions of wild and unacceptable femininity.

The ancient Greeks saw her as the daughter of Chaos and Darkness, the primal creative force and the void into which it manifests. They called her a Goddess of Chaos, Discord and Confusion, for in their rigid social paradigm they could not conceive that one must have Chaos in one's "soul" to give birth to a dancing star, that confusion seems a holy state entirely preferable to all fragile certainties, and that discord seems the precursor of all great synthesis. The universe proceeds by confusion, indeterminacy and trial and error, otherwise how could creation and evolution occur? Or how could we have occurred?

The invocation of Eris marks a celebration of paradox, of contradiction, of fertile confusion, of that which lies beyond logic, of those qualities repressed by patriarchal regimes and unfairly relegated to the feminine domain, emotion, feeling, socially unaccepted liaisons, hysteria, enthusiasm and untidiness.

A Priestess most commonly performs The Mass of Chaos E, although a Priest may take the role if he dares.

During the cacophony participants shout out any system of belief or beliefs they wish, choosing from real, imaginary, or nonsensical systems or all of these things simultaneously, they also may argue two or more beliefs against each other with themselves, particularly in a solo invocation.

Substantial quantities of screwed up newspaper can mark the circle, and participants should freely tramp through this during their circumambulations.

The Priestess should memorise the Enochian Incantation and the Vernacular Proclamation and indeed the general format of the rite beforehand rather than read it from papers. The Priestess brandishes an apple with a K carved into it during the ritual, although some other fruit may serve in an emergency.

The Priestess consecrates the apple with a kiss and/or by taking a bite from it. She may then drop it into a chalice of fruity liquid to make a sacrament, or perhaps crush it on to a sigil or pentacle for some magical purpose in a solo ritual. Afterwards she may count the seeds for divinatory purposes, plant them or retain them as talismans.

THE MASS OF CHAOS (E)
The Rite:

Statement of Intent:
Participants stand round perimeter of circle

WE WILL TO INVOKE ERIS TO CONSECRATE THIS SACRAMENT

TO CHAOS FOR OUR MAGICAL INSPIRATION (Or other related intent at will).

2. The Priestess draws the Sigil of Chaos above the circle, all visualise.

3. The Priestess shouts KALLIATI! to begin the cacophony.

4. The Cacophony: Participants circumambulate widdershins loudly and passionately arguing and proclaiming various beliefs.

5. The Priestess gradually dances a spiral inward to the center of the circle and there shouts KALLISTI! to end the cacaphony.

6. The circle dance continues and participants begin chanting somewhat more quietly ERIS, ERIS, ERIS, randomly interspersing this with equivalent goddess names such as ISHTAR, BABALON, LILITH, and KALI. Participants point towards the Priestess with the left hand and visualise the Sigil of Eris into her. The Priestess meanwhile delivers the Enochian incantation to Eris, as many times as felt necessary.

7. The Priestess makes the Vernacular Proclamation, chanting ceases, but visualisation continues.

8. The Priestess consecrates the fruit, and serves the sacrament if used.

(8.5 Erotic Gnosis may follow if desired and previously agreed, though this usually remains confined to workings with small numbers of participants).

9. Laughter Banishing closes the rite.

The Enochian Incantation:
OL NIIS VOHIM OLGIZYAX
I come with a hundred mighty earthquakes

ZIXLAY DODSEX QAAON
To stir up vexation in your creation

OHORELA QAAS NETAAIB CAOSAGON
I made a law of your creation for the government of the earth

OVCHO CORDZIZ EOLIS OLLAG ORSABA
Let it confound those who reason, making men drunk

AGLO I VAOAN OD TOLGLO NOALN
Nothing has truth and all things can happen

ZIRDO PASAB DE DOSIG OD TEHOM QUADMONAH
I am the daughter of night and the primal chaos

ZIRDO IADNAMAD CYNIXIR MOZAD FABOAN
I am the undefiled wisdom, mingled with joy and poison

MICAMA ISRO QUASHI
Behold the promise of pleasure

NANAEL IPAMIS ADIRPAN

You cannot cast my power down

ZODACAREE ZYLNAR NANAEL ZODAMERANU
Move! in itself my power, Phenomenize!

IXOMAXIP KALLISTI IO ERIS!
Let us know her, the fairest, Hail Eris

The Vernacular Proclamation of Eris

Aeons ago I offered you this apple

But you built religions, laws and prisons for the mind
You must have chaos in yourself to give birth to a dancing
star

I bring the inspiration from which your artists & scientists
build rhythms.

I bring the anarchic laughter of children of all ages,
I bring Chaos and I bring life

In this new aeon I come before you as Eris
To tell you to seize freedom,

Nothing has absolute truth, Anything can happen!
The World lies in my Apple! (Administers Kiss and/or
Bite).

Appendix 3

Sorcerer's Chess.

Sorcerer's Chess takes place on a conventional chessboard with conventional pieces and all the conventional rules, but it also includes extra functions for four of the pieces and some additional rules.

The game closely resembles Discworld's Stealth Chess in that it resumes the quantum mechanical principle of indeterminacy in the actual position of certain pieces.

However it has a fully self consistent set of rules and it does not require the considerable feats of memory demanded by its Discworld variant.

Each side has two Pawns designated as 'Assassin' pieces and these should each have a rotating dial fitted visible near the base showing the numbers $0 - 7$. These Assassin pieces start the game on the squares of the Rooks's Pawns, with their dials set to 0.

In each turn a player may move an assassin piece instead of moving a conventional piece. The Assassin piece may make a 'Real' move of one square per turn in any direction or it may make a 'Virtual' move in which case it remains in position and has the number on its dial increased by one, up to a maximum of seven.

Additionally the Assassin piece may make a 'Revealing' move instead, in which case it moves directly from its square on the board to any square within range of the number shown on its dial. When it makes a Revealing move the player resets its dial to 0.

When making a Revealing move an Assassin piece captures any piece on the square it lands in, except a King, whose square it cannot land on. It eliminates pieces from its own side by this action also, should a player find such a strategy advantageous. Note that an assassin piece can nevertheless threaten a King if it lands or moves next to it in 'real' space.

Whilst an Assassin piece has anything other than zero showing on its dial it doesn't actually 'occupy' the square on which it stands and it cannot move in real space, players should regard it as occupying some sort of virtual space with its position on the board merely marking its point of departure.

Thus an Assassin piece in virtual space can occupy the same square as another piece, it does not obstruct the movement of other pieces and the opponent cannot capture it. However an Assassin piece in real space, with its dial at zero, behaves like a normal piece.

Assassin pieces for Sorcerer's Chess

These rules make the Assassin pieces powerful but not necessarily dominant in the game. Whilst they can launch surprise attacks they also risk making themselves venerable to attack after they do so, particularly to other assassins, players should try to cover all pieces as in ordinary chess.

Chess itself resembles a semi-abstract medieval conflict, so the Assassins in Sorcerer's Chess represent characters creeping around a castle unseen in the wall cavities and floorboard spaces or posing as harmless servants.

In a more esoteric sense the Assassins' moves represent the effects of quantum physics with real moves and the occupancy of real spaces representing particle mode and virtual moves and virtual space representing wave mode. Thus the Assassin pieces function rather like a cross between a ninja and a magician.

On an historical note we observe that modern chess developed from the ancient Indian game of Shaturanga which originally had four players. The condensation of Shaturanga into a two-player game led to the duplication of the pieces now known as rooks, knights, and bishops. The wizards of the Golden Dawn devised a somewhat ungainly adaptation of Shaturanga played with figurines of Egyptian gods which they called Enochian Chess because of the tablets derived from the Elizabethan wizard Dee, on which they played it.

Although ripe, and perhaps over-ripe with arcane symbolism, Enochian Chess did sometimes serve as a divinatory tool.

Nevertheless one can trace a simple alchemical symbolism present in Shaturanga to the modern form of chess. Rooks to earth, knights to fire, bishops to water, queens to air, and kings to spirit, and thus perhaps in sorcerer's chess, assassins to the astral.

Appendix 4
Symbols and Constants

1) Physical Constants.
(MKS units, metre, kilogram, second)

Lightspeed c,
~3 x 10^8 m/sec

Planck's Constant reduced $\hbar$,
~1 x 10^{-34} joule seconds.

Gravitational Constant G,
~ 6.67 × 10^{-11} m^{3}kg^{-1} sec^{-2}

Boltzmann's Constant k,
~1.38 x 10^{-23} joule/kelvin

Entropy S,
Measured in joules/kelvin

Absolute Temperature K^0,
Degrees Celsius + 273. $0K^0$ = absolute zero.

Ubiquity Constant U,
~6 x 10^{60}

Anderson Deceleration A,
8.74 x 10^{-10} m/sec^2

2) Size of Universe. (Based on A)

Mass
1.38×10^{53} kg

Antipode Distance
1.03×10^{26} m, ~11 billion light years

Temporal horizon
3.34×10^{17} sec, ~11 billion years

3) Magical units

Roundworld
Prime Π,
$\sim 10^{-14}$ joules/sec

Roundworld
Thaum Θ,
$\sim 10^{7}$ bits/sec

(Discworld Prime and Thaum have values about 50 times higher)

4) Miscellaneous symbols
Amount of magic brought to bear Ψ, equivalent to 1 Prime or 1 Thaum when $\Psi = 1$.

Integrated information Φ, inversely proportional to entropy, in numerical units of S/k

Indeterminacy Δ, the undecidedness in the value of a quantity.

References etc.

1. Aeon. – A word used here to denote a phase in the development of metaphysical thoughts and paradigms. See Liber Kaos Chapter 2, for the Psycho-historic mechanism of the aeons.

2. Abramelin. - A famous 15th century Grimoire. *The Book of the Sacred Magic of Abramelin the Mage* translated by S.L. MacGregor Mathers (1897; reprinted by Dover Publications, 1975)

3. Phlogiston. – A soon discredited 17th century hypothesis that supposed that flammable substances owed their flammability to the presence of this substance.

4. Poincare, Jules Henri. 1854 – 1912 French mathematician, theoretical physicist and philosopher of science.

5. Grigori Perelman. – Russian mathematician born 1966. Credited with proof of the Poincare conjecture in 2006. Declines all prizes, lives reclusively.

6. Anderson, John D. Leader of the group that discovered the Pioneer Anomaly. John D. Anderson, Philip A. Laing, Eunice L. Lau, Anthony S. Liu, Michael Martin Nieto, Slava G. Turyshev (1998). "Indication, from Pioneer 10/11, Galileo, and Ulysses Data, of an Apparent Anomalous, Weak, Long-Range Acceleration". *Phys. Rev. Lett.* 81: 2858–2861.

7. Gödel, Kurt.1906 – 1978. Austrian logician. Famous for his Incompleteness theorem.

8. Heraclitus of Ephesus. – Iconoclastic and contrarian Greek philosopher of the 6th century BC. His riddles and strange pronouncements continue to intrigue.

9.The Left Hand of Creation. John D. Barrow & Joseph Silk. ISBN-10: 0195086759

10. The God Delusion. Bantam Press 2006, by Richard Dawkins ISBN-10: 0593055489

11. Josef Gobbels. 1897 – 1945. Nazi Propaganda Minister.

12. Eliphas Levi. 1810 – 1875. French occultist author and magician.
Dogme et Rituel de la Haute Magie, (Transcendental Magic, its Doctrine and Ritual), 1855 Histoire de la Magie, (The History of Magic), 1860

13. Principia Discordia. – Malaclypse the Younger. Still available from a few questionable sources.

14. Ouranos audio CD. - Original Falcon Press. http://www.originalfalcon.com/

15. Cthulhu Mythos. - Also known as the Lovecraft Mythos, created in the 1920s by American horror writer H. P. Lovecraft, with additions from his contemporaries.

16. Boltzmann-Gibbs. - Ludwig Boltzmann came up with the original concept of entropy. Josiah Willard Gibbs refined it.

17. Shannon, Claude Elwood. 1916 – 2001. American mathematician and electronic engineer, known as the father of information theory.

18. Beckenstein-Hawking. Stephen Hawking & Jacob Bekenstein conjectures on Black Hole Thermodynamics.
Hawking, Stephen W. (1974). "Black hole explosions?". *Nature* 248 (5443): 30–31.
Hawking, Stephen W. (1975). "Particle creation by black holes". *Communications in Mathematical Physics* 43 (3): 199–220.

19. Heisenberg, Werner. - 1901 –1976. German theoretical physicist who discovered the fundamental quantum uncertainty relationship between position and momentum. An insight later generalized to a whole suite of seemingly complementary and indeterminate quantum values. Hence the classic graffiti slogan, 'Heisenberg was here – probably'

20. Minkowski, Hermann. - 1864 – 1909. German mathematician, one of Einstein's teachers, a pioneer of higher dimensional geometry.

21. Sheldrake, Rupert. - A New Science of Life: the hypothesis of formative causation, 1981 (second edition 1985). ISBN 0-87477-459-4.

22. Emmy Noether. 1882 – 1935. Jewish- German-born mathematician. Her revolutionary theorem explains the fundamental connection between symmetry and conservation laws.

23. Gnosis. – A word originally denoting knowledge or wisdom, but in modern Chaoist parlance it denotes the state of heightened or ecstatic awareness itself rather than any particular facet of knowledge.

24. Groundsleves. - A Chaos magic term for the material basis of an evoked entity. Typically groundsleves consist of portable figurines and fetish objects.

25. Nucleosynthesis. – The process in which stars transmute hydrogen into heavier elements. See The Magic Furnace, Markus Chown, ISBN-10: 0099578018.

26. Tononi , Giulio. Consciousness as Integrated Information: a Provisional Manifesto. *Biol. Bull.* 215: 216-242. (December 2008)

Phenomenise. We apologise for this neologism. By its use we attempt to circumvent or at least draw attention to the ridiculous grammatical implication that events have some 'being' separate from their actual manifestation.

Books by the author.
Liber Null & Psychonaut. Pub. Samuel Weiser Inc 1987. ISBN 0877286396

Liber Kaos. Pub. Samuel Weiser Inc 1992. ISBN 0877287422

Psybermagick. 3rd edition, Original Falcon Press 2008. ISBN 978193510657

The Apophenion. Pub. Mandrake of Oxford 2008.
All still in print at the time of writing.

Index

Art prints of Matt Kaybryn's paintings
created for 'The Octavo' are available
from the artist's website:

w w w . v i s u a l g n o s i s . c o m

VisualGnosis

Perceptible by the mind:
knowledge and insight
into the finite and
unbounded, divine and
uncreated in all,
under and above all.

Mandrake

'Books you don't see everyday'

The Apophenion:

A Chaos Magic Paradigm by Peter J Carroll.
978-1869928-421, £10.99

My final Magnum Opus if its ideas remain unfalsified within my lifetime, otherwise its back to the drawing board. Yet I've tried to keep it as short and simple as possible, it consists of eight fairly brief and terse chapters and five appendices.

It attacks most of the great questions of being, free will, consciousness, meaning, the nature of mind, and humanity's place in the cosmos, from a magical perspective. Some of the conclusions seem to challenge many of the deeply held assumptions that our culture has taught us, so brace yourself for the paradigm crash and look for the jewels revealed in the wreckage.

This book contains something to offend everyone; enough science to upset the magicians, enough magic to upset the scientists, and enough blasphemy to upset most trancendentalists.

Order direct from

Mandrake of Oxford
PO Box 250, Oxford, OX1 1AP (UK)
Phone: 01865 243671
(for credit card sales)
Prices include economy postage
Visit our web site
Secure online at - www.mandrake.uk.net
Email: mandrake@mandrake.uk.net

CPSIA information can be obtained
at www.ICGtesting.com
Printed in the USA
BVHW080744270620
582373BV00002B/260